"A timely and invaluable contribution to this culture...at the cutting edge of the challenges modern people face everyday."

—Malidoma Somé, author of *Ritual: Power, Healing and Community*

"This insightful book documents the terrain of our touch-starved culture and gives concrete suggestions on how to...be more in touch with ourselves, and therefore more in touch with others. As technology becomes a more pervasive part of our lives, the need for touch becomes greater than ever."

—Chistiane Northrup, M.D.

"A significant contribution to the literature on touching in human life and relationships. *Untouched* brings unusual insight and wisdom to understanding the nature of human and sexual love. Highly recommended for those who want to be 'in touch.'"

—James W. Prescott, Ph.D. Director, Institute of Humanistic Studies

UNTOUCHED

THE NEED FOR GENUINE AFFECTION IN AN IMPERSONAL WORLD

MARIANA CAPLAN, M.A.

HOHM PRESS

Layout and Design: Visual Perspectives, Phoenix, AZ
Cover Art: Kim Johansen

Library of Congress Cataloguing in Publication Data:
Caplan, Mariana, 1969-
Untouched: the need for genuine affection in an impersonal world / Mariana Caplan.
p. cm.
Includes bibliographical references and index.
ISBN 0-934252-80-7 (pbk. : alk. paper)
1. Touch--Psychological aspects. 2. Touch--Social aspects.
I. Title.
BF275.C37 1998 97-48934
158.2--dc21 CIP

Hohm Press
P.O. Box 2501 • Prescott, AZ 86302
(800) 381-2700 • Fax (520) 717-1779
E-mail: pinedr@goodnet.com
Web Site: http://www.booknotes.com/hohm/

I am not a mechanism, an assembly of various sections.
And it is not because the mechanism is working wrongly,
  that I am ill.
I am ill because of wounds to the soul, to the deep emotional
  self
and the wounds to the soul take a long, long time, only time
  can help
and patience, and a certain difficult repentance
long, difficult repentance, realization of life's mistake, and
  the freeing oneself
from the endless repetition of the mistake
which mankind at large has chosen to sanctify.

– D.H. Lawrence, *Healing*

ACKNOWLEDGMENTS

My acknowledgments go to the forefathers and foremothers who brought the message of the vital necessity of touch, affection and intimacy to the forefront of public attention: Ashley Montagu, Joseph Chilton Pearce, James Prescott, and Jean Leidloff.

Also, my thanks to the reference librarians at the Prescott Public Library who spent countless hours anonymously assisting me with research.

And, as always, my gratitude belongs to Lee, my teacher, my motivation, my inspiration, and my Friend.

DEDICATION

This book is dedicated to all the children who didn't get what they needed, and to the possibility for this to change in future generations.

C O N T E N T S

FOREWORD
by Ashley Montagu

Mariana Caplan has written a beautiful book, wonderful, courageous, and distinguished by its eloquence, soundness and originality, a book that all the world should read, for it carries a healing message, easy to apply, full of highly realizable promises, and happily rewarding, especially for those of us who have been failed during the earlier years of our lives, and who bear the scars of that trying misadventure called "bringing up" or "raising" the child. I have for long held that adults are nothing more nor less than deteriorated babies, "adulterated" by caregivers unaware of their own victimization who, more often than not, with the best of intentions were zealously devoted to doing their best for their charges. Not only that, the professional authorities, obstetricians, pediatricians, psychiatrists, psychologists, psychoanalysts, psychotherapists, educators, and other non-swimmers acting as lifeguards, have for the most part perpetuated the errors of their predecessors. With their superficial psychologies they have overlooked the essential fundamental biosocial requirements for the healthy growth and development of the person. Let me say at once that by healthy growth and development I mean the ability to love, to work, to play, and to use one's mind soundly. But these are words, and they all require definition and proof. (I have done this in my book, *Growing Young*, 2nd ed. Westbrook, CT: Bergin & Garvey, 1989.)

Words shape our understanding, and definitions can only be soundly meaningful at the end of an inquiry rather than at the beginning of one. It is the *tested* evidence that establishes its validity. It is by such means, by experiment, that a scientist comes to believe in proof without certainty, whereas others believe in certainty without proof. Most of us tend to think in stereotypes, in clichés, in fixed conventional expressions, and so we communicate with phrases and words as if the presuppositions upon which they were based never existed. Take any word at random and ask anyone to define what they understand by it, and see what you get. You will find that in most instances people are unable to give an accurate account of the words they commonly use. Ask your friends to define a definition. Don't be surprised that they can't do it. And so it goes. Ask your students at high school or university level to define the meaning of any of the words they customarily use. Ask them to define the word "race." The answer they will give you will almost certainly be wrong, for in spite of the findings of anthropologists, biologists, and geneticists that the term "race" corresponds to nothing on earth, especially as it relates to human beings, but constitutes a social construct created by scientists during the 19th century in the endeavor to account for the perceptible differences between different peoples. The result was an elaborate "authoritative" typology and classification that led to the torture and extermination of millions of human beings. Most people, however, continue to believe that "race" is a very real thing. Plato many years ago believed that the word created the thing. He was, of course, right. As Aldous Huxley pointed out, the overvaluation of words leads all to frequently to the fabrication and idolatrous worship of dogmas. Hence the importance of education.

Unfortunately, what we understand by "education," isn't education at all: what it is, is *instruction,* a very different thing, a training by *rote* in *what* to think, not in *how* to think. Hence, it is not surprising that most of us find it easier to believe than to think, to have faith rather than to examine the presuppositions upon which our beliefs are based.

It was Plato who long ago said that the unexamined life was not worth living. How right he was, and how right Mariana Caplan is, in this splendid, most readable, and important book, in which she so congenially helps us all to reexamine the basic beliefs which normally form our daily round of living. Beyond all else, as she so illuminatingly shows, there is the crying need for affection that, in a world of untouchability, is so seldom understood. This is the setting and the challenge that is dealt with by Mariana as it has never been so satisfactorily and effectively dealt with before, the behavior, the sensibility, with which we are touched into completeness. IT is that need for love and to love others, which takes physical expression through touch, in a hug, an embrace, a touch of the hand, a kiss, and more, that constitutes our principal demonstration of connectedness, involvement, and caring. Indeed touch becomes the touchstone, the true measure of our humanity, our seasoning of health, the enrichment of our being. Hence, Mariana discusses, as they have never been so creatively discussed, virtually every significant human need and behavior as it has never been so revealingly examined, analyzed, and creatively restored to understanding and usefulness, in language that abjures all technical terms, and speaks plainly and simply, both to the heart and the mind's consent. This is a considerable achievement. Enthusiasm is good, but it should not look to long at the sun, and keep others from sharing the light, the glow, of warmth that this extraordinary book communicates.

Ashley Montagu

INTRODUCTION

You could say I had a "normal" childhood—normal in a time when earthquakes and global warming don't even make the front page; normal in a time when homicides and suicides are a daily occurrence and the threat of nuclear war is run-of-the-mill; normal in a time when it is dangerous for children to play in the streets in once-safe neighborhoods, and for women to go alone almost anywhere, anytime; normal in a time in which people do not know how to extend genuine love and respect to one another, when it is common to go for days or weeks with no physical contact with another person; normal in a time when we have all but lost our humanness.

I had a "normal" happy childhood (just as all of my therapy clients, and most ordinary people, claim to have had "normal" happy childhoods). Raised in a normal dysfunctional family (with some rare exceptions, "dysfunctional =

family" in contemporary Western culture), there were expectations and rules regarding touch that were to be observed. For example, when your parent says, "Come sit on your Daddy's lap," you sit on your Daddy's lap, irrespective of your wishes as a child. When your uncle squeezes you too tightly, or slobbers all over your face when he kisses you, you endure it politely. When somebody wants to pinch your cute little cheeks, or pat your head, you force a smile. You are a child, and therefore at the whim of the unmet needs for affection that others randomly place upon you.

We were a normal family in the pre-computer era. Yes, atomic bombs were dropped at Hiroshima and Nagasaki. Sure, there was a nuclear reactor meltdown at Chernobyl, releasing a radioactive cloud that contaminated 2,000 square miles of the earth. Like all my friends, I lived through a couple of nuclear war threats, and watched the news when escaped convicts were running around my neighborhood. Yes, my innocent body was fed TV dinners and microwaved food, and I watched *General Hospital* and longed to be anorexic so I could be skinny enough—but still, I maintain, this is *normal*. I was no different from any other kid on the block. We were a happy American family just like the rest of them—dead, you could say, but at least we were normal.

Such a scenario is all too common. As evidenced by the current skyrocketing statistics on child abuse (averaging between thirty and seventy percent depending upon your source), it is shockingly common for children to be severely damaged by the weapon of touch. Whereas some perpetrators of touch crimes (often the parents themselves) are blind to the lifelong consequences that their actions will have on the child, others are simply unwilling to place the child's needs over their own aggressive impulses.

Yet to place the entirety of the blame on any single parent's shoulders is to bypass the core of the problem. For parents are simply children who have grown up in what I will refer to as a "touch-starved nation." They, too, have been raised in a culture that has lost its roots, and during a time when people are so excited by science and technology that they are all too willing to be seduced away from their essential integrity, sanity and traditional family values in exchange for the allure of television, fancy cars, an important job, a computer *with* Internet and a CD Rom, and the mindset that accompanies this. They, too, are simply the products of a growing trend that, by design, will increasingly alienate us from ourselves and therefore from others.

All children instinctually understand sanity before it is drained out of them. Tens of thousands of young children look out at the world through the safety of their innocence, at their parents drinking or fighting, sitting zoned out in front of the television or the computer, stuffing their hungry hearts with one addiction or another. From a very young age they witness abuse and violence—in the papers, on the television, in their own homes. Given today's world, it is nearly impossible to raise a secure, happy, healthy child who is untainted by the ills of contemporary culture. Even if well-intentioned parents manage to create a loving home; even if their children's bodies do not become overly toxified by the chemicals in the food and water that they ingest, or the radiation emissions from their television sets, these children are still going to be faced with the toxic and often cruel world that is just outside their doorstep.

My point is that touch-starvation is not an affliction of the few, but is spiraling outward into the culture of the masses. The awkwardness and inappropriateness of where, how and whom we touch is an issue that affects *everybody.* Whether you are somebody who is unaware that you even *need* touch, or a single man or woman who has been horny for the last five or fifteen years because you are not in an intimate relationship and the only way you know how to get touch is through sex, or even if you are surrounded by family and children who want to love you and touch you but you don't know how to take it—you *are* affected by touch, or the lack of it. Who does not at times question whether it is O.K. to put their arm around their teenage child or co-worker, or to initiate sex with their spouse, boyfriend or girlfriend, or how to say "no" to unsolicited and unwanted touch?

In my normal happy childhood, I was fortunate enough to be exposed to positive forms of touch as well. Beginning in my teenage years, with an intuitive awareness that there had to be an alternative to the middle-class, sheltered civilization of deadness to which I had belonged all my life, I began to seek out any excuse I could find to spend extensive periods of time in foreign countries. My first expeditions were spent with families in Colombia, South America and in small villages in the Yucatan Peninsula in Mexico. The seeming inconveniences of seven people living in one room, or sleeping in a hammock and taking cold showers out of a bucket, were negligible compared to what it felt like to live in an environment of intimacy, communion, love, and an overall sense of well being. In the effort to quench my enormous thirst for understanding, I pursued a career as a professional traveler, becoming a cultural anthropologist and immersing myself in cultures that were both full of, as well as devoid of, touch. Still dissatisfied with the education my heart was receiving, I turned to the study of psychology—delving into an investigation of both my own psyche, as well as the psyches of my clients. Yet my knowledge remained incomplete. Much work was occurring on the level of mind, but my clients would continue to come in with body postures of pride, insecurity or terror. Thus, I sought balance by studying a particular form of bodywork that gave specific attention to the relationship between the body and the emotions.

Both at home and out of the country, I began to immerse myself in environments in which people were attempting to live another way, finally finding myself in my present community environment—a place in which touch and the need for touch is carefully and consciously considered. For example, in our sub-culture, children are never hit, and are instead given plenty of safe and loving touch according to their own wishes and needs. There is also a sensitivity to the needs of adults who suffer the wound of unlove, and attention is given to providing them with supportive touch in their adulthood, in the hope of allowing even part of the missed process of bonding that should have happened in early life to take place. My travels, adventures and experiments in community have been far-reaching and at times radical, but not because I am unusual or eccentric. Normal just like everybody else, I am driven by an instinctual knowledge that there is another way—a way of greater sanity.

I do not claim to be the holder of this "other way." This book does not lay out a revolutionary plan as to how a contemporary Western culture full of touch and intimacy and generosity and caring will look. Instead, it points in a specific direction, and it is for us to look that way to discover just how that touch will show up within ourselves, our families and our communities.

This book includes many examples of other cultures who seem to have an inherent understanding of the importance of touch. The purpose of this is not to idealize these cultures, nor to encourage people to become Third World wannabes—for these cultures have their own problems. We don't have to start dressing in bright African gowns, or taking our prayers out of the chapel and into the Native American sweat lodge in order to rediscover touch in our lives. We live in the West—whether it be America, Canada, France or England—and we are products of the Western world. There is not a chance that families will sleep together on woven palm mats, or that mothers will strap their babies onto their backs while they go out into the rice fields, or that extended families including older siblings, grandparents, aunts, and neighbors alike will all assume responsibility in caring for a young infant.

However, we can integrate the best of the *essence* of other cultures into our own. We have certainly not hesitated in integrating ourselves into the politics, economy, and customs of other cultures—as evidenced by the Domino's Pizza sign painted on the side of a building in a Mexican village where many cannot even afford to eat; by "U.S. Army" camouflage gear being sold outside of an ancient temple in southern India; by a growing colonization of Western ideals throughout the world. If we intend to maintain any semblance of balance in our world, we must also be willing to embrace and include in our own culture the best qualities of those cultures that we have come to dominate.

The values I refer to are not indigenous only to foreign cultures. They once belonged to our own culture and are aspects of all cultures that live even remotely close to the earth, cultures in which the essential aspects of community are still intact.

The issue of touch in the present day is not a question of blame. In an effort to deepen our understanding, the origins of this issue will be examined and the perpetrators named. This is not to blame any one individual. When we come to understand the gravity of the situation, and how it wreaks havoc on us moment to moment, we know that there is no value in pointing a finger, except perhaps at ourselves. For although we are the victims of touch crimes, we are also the perpetrators. We are also the only ones who have the capacity to create change. This is just what I am proposing that we consider together.

The very small gestures we begin to make, as a result of a clearer understanding of the lack of touch in our lives, are received by others as marked and significant. Though we cannot heal the soul-wound by an affectionate pat on the back, for those who have felt unloved all of their lives, a small act of kindness can shake their whole perspective about who they are in the world. Touch, when done with heart, is always healing—period. Whether given by a trained nurse or a nervous friend, it heals.

On the most basic level, I will consider the need for simple, well-intentioned, physical touch. Everybody, and particularly young children whose innocence is still intact, needs this touch. However, touch is much more that skin contacting skin. Touching has to do with the acknowledgment of our shared humanness. It has to do with the recognition of the inherent vulnerability and intense wish for contact that is present in each of us. Touching results from an acceptance of the separateness of each individual, and the knowledge that it is only through contact that union and communion can come about. Beyond this, the possibility of fulfillment through the medium of touch—whether it be physical touch or not—is far beyond what most people have ever known or considered. And this possibility belongs to all of us.

When people feel loved as a result of the abundance of touch and affection in their lives, they naturally extend themselves to touch others—be it by a simple pat on the shoulder or a touch on the hand. Secure in themselves, they are open to the other's response, but not expecting it. Their sense of safety and inner stability does not depend upon how other people respond to them. They touch in order to express themselves, and so that others may feel cared for. They wake up in the morning feeling loved, and go to sleep feeling loved—no matter what particular circumstances arise in any given day. From this love, which is the result of genuine bonding, arises wholeness, commitment, loyalty, service.

Throughout the course of this book, I set forth a detailed description of the degree to which we as a race have lost touch. I am unabashed and uncompromising in my opinions about the pervasiveness of all forms of abuse, the terrifying ways that we birth our children into this increasingly maddening world and raise them to become unfeeling machines, and the dangerous state that we as a people are in.

The actual physical touch—if it is True Touch—is the end of the road. That is to say that by the time somebody is able to give or receive affection from another person, deep understanding and healing have already taken place. Just as one must look to the sky to fully understand why the ocean is blue, so this book addresses touch by revealing the cause, the foundation and the context of touch. This book does not explain "how to" touch, but instead unfolds *the way* of touch. The human being whose mind state has been revealed to him clearly, and who has understood the roots of his own suffering, will naturally and necessarily desire to change and to transform, the "how to" emerging of its own accord and becoming refined through a process of trial and error.

There are thousands of parents, massage therapists, health practitioners, teachers, priests and lovers who respect the value of touch. They consciously practice it because they feel the truth of it, yet many do not know why they touch, or how they know to touch. There are also many more hundreds of thousands who do not touch, do not think to touch, do not want to be touched. Similarly, most of these people do not know why they don't touch, and many believe they do not know how to touch, or that they are inherently not inclined to touch. This book takes the reader behind touch, behind the skin, behind the urge to touch or not to touch.

You, the reader, already know the underlying principles behind everything that you will read in this book. I write from the premise that the knowledge of all things that are essential for living a sane, balanced and loving life (as is touch, as is love) is alive in everyone, only it has been covered by years of conditioning, abuse, falsity and a cultural mind that is dictated by fear. The gold is in everyone, only many are not willing to go digging in the mines.

I would like to extend my empathy for the pain that will be felt by those who do not take offense at my words and instead allow themselves, perhaps for the first time, to be faced with the reality and the gravity of our present situation. If you allow me to, I will walk you through a shattering of an illusion—i.e., that the human race is progressing in terms of its understanding, and that people are living according to what it means to be human. It is my sincere wish that from this shattering will arise something that is genuine and whole—a way of being in the world that allows you to feel good about yourself and to know that you are fulfilling your rightful place.

Something happens in people when they hear what they know to be true, and something still deeper occurs when they take steps toward living that truth. The enormity of the challenges and heartbreak that often accompany facing one's reality without buffers is subsumed by the sense of "rightness" and integrity one finds within oneself and in one's life. It is my wish to draw this forth from you, the reader—the knowledge of what it means to really touch another, and the courage to throw down the reins and risk living it. For if we who care enough to consider this possibility do not allow ourselves to recognize this truth and to live accordingly, who will?

CHAPTER 1

A Touched-Starved Nation

In our crowded and urban world, we have battled on in this way, further and further from a state of loving, personal intimacy, until the cracks have begun to show. Then, sucking our metaphorical thumbs and mouthing sophisticated philosophies to convince ourselves that all is well, we try to sit it out. We laugh at educated adults who pay large sums to go and play childish games of touch and hug in scientific institutes, and we fail to see the signs. How much easier it would all be if we could accept the fact that tender loving is not a weakly thing, only for infants and young lovers, if we could release our feelings, and indulge ourselves in an occasional, and magical, return to intimacy.[1]

– Ashley Montagu

If you have ever looked closely at the barnacles on the rocks at the edge of the sea, you will notice that they appear as thousands of tiny mouths that are attempting to eat, eat, eat...No matter what you place in front of them, their "mouths" continue to reach out for food. In this same way the skin and heart of the vast majority of people in the Western world are hungering for touch. We are famished for the most basic needs for human contact and closeness. We are living with a feeling of deep disconnection and fear. We sense that something has gone very wrong, but we have no idea what. We are dying and only vaguely conscious enough to know it. The madness we have created in our lives and in our world is our way of crying out to some imagined savior, just as the hungry infant will scream and cry until he is fed. We are the walking wounded. We are the out of touch. We are the touch-starved, love-starved products of the material world. Our very lives are the casualties of the modern way of television and computers, and we are the current survivors in the age of nuclear threat. We are both the children of parents who did not know how to love, as well as the struggling parents who are trying to raise our children sanely in an insane world. We are it—the starving attempting to feed the hungry, the wounded attempting to heal the sick. We are the ignorant looking for answers, the blind attempting to live out a vision. We are the hope in the hopeless situation, the possibility attempting the impossible.

ON COOKING FROGS

If you place a frog in a pot of boiling water, it will immediately jump out. However, if you place a frog in a pot of cold water, and then very slowly heat the water to a boil, you will end up with a dead, cooked frog.

I do not mean to be insulting by reducing the human condition to that of cooking frogs, but it is a precise parallel to the way in which present day society is slowly taking the life out of us, leaving us feeling increasingly deadened without even knowing why.

"But how can this be?" we want to know. "We are just going along with the times." "We are living in an age of unprecedented progress and prosperity." "We are just sitting in a pot of cool water with the other frogs enjoying the abundance."

The price of the so-called progress we have made is our present state. Of course we all say we're doing "fine," but what does it really mean to be doing fine? Yes, we go out to dinner, go on vacations, pursue interesting hobbies, get married, have children and live in nice houses with nice things if we're lucky, but all of these activities are simply measures of survival—frantic endeavors to put bandages over our collective wound in an effort to avoid the full impact of the underlying depression and fear that we all feel; an attempt to somehow convince ourselves that living in a pot of boiling water is not dangerous.

When I looked for an example of healthy touch within contemporary culture as a model for this book—some group of people who were really *in touch* with themselves and their surroundings—I was unable to find one. Where were the people who could touch one another frequently and easily? For whom intimacy was natural, and who felt good about themselves? Where were the people who were genuinely content with their lives? (Not just a compromised position of, "Looking around, I seem to be doing as well as anybody.") What could I give my readers to relate to? To aspire to? I mentioned this problem to a friend of mine who responded by saying, "That's why you're writing the book."

I am not saying that there are absolutely no happy people in the Western world—land of the "We're in control and everything is fine." Of course there are some, and we all know who they are when we meet them. However, if we have to think about whether or not we ourselves or somebody we know is truly happy, we, or they, are not. "I guess I'm happy" is not happy. "He has a good job and a nice family, of course he's happy," is not happy. "I've got it all," is not happy. Such is the frog as it swims around in the small, warm pond on the stove, assuming that everything is fine. Perhaps it has a slight instinct that something is going wrong, but since it is in the company of other dying frogs, it is far too wrapped up in mating, having babies, and getting food to consider the possibility that it is being cooked alive. What would make it think that it was dying anyway?

It is not that somebody or something is specifically trying to kill us. It is more complex than that. This momentum is being driven by a combination of forces that are created from a context of fear, and are primarily motivated by a wish to gain power, money and control. A set of values has been locked into place that give precedence to thought over feeling, independence over connectedness and service, and quantity over quality. In order to sustain this wave of commercialism, consumerism and technological advancement, and to allow its momentum to increase still further, we must consume more, buy more, spend more, eat more. Much as the addict depends on his substance of choice, this momentum has become our cultural "substance of choice." We feed off of it, and it off of us. We find a kind of nourishment in it. It does not feed the soul, but it does feed the hole of loneliness and emptiness that so many people feel within themselves. That hole is fed by anything sugar-coated, caffeine-coated, stimulation-coated, get-me-away-from-reality-coated. This wave of consumerism provides us with an unending supply of numbing agents: gadgets, movies, foods, discoveries, escapes...all designed to ward off feeling. We nourish this wave in turn by keeping ourselves ignorant of it and at the effect of its force.

Fighting the wave of commercialism and consumerism is a very complex situation, really. It is as if this anti-life force has become so strong, so sophisticated and so seductive that it has swept us up in its wake. We must not underestimate, nor be naive about the power of this force. We are not unintelligent, but when everything in and around us—every advertisement, every emotion, every teacher, every politician, and every ideology is saying one thing...when all of this is washing across us with an unprecedented degree of uprooting force, it is nearly impossible to not get carried away by it. Standing clear in one's own knowledge in the face of this force can be likened to trying to swim out to sea while a tidal wave is heading to the shore. Who would try, unless they knew that the only option was sink or swim?

We would like to think that we are too smart to be fooled. In fact, we are fully capable of knowing exactly what is going on, only we do not want to know, as it confronts us with too great a responsibility. Consider the millions of Germans who sat back and did nothing as six million Jews, gays, gypsies and artists were murdered. The Germans were not stupid. They knew who was taking their neighbors away and that they would not be seeing them again. Daniel Holdhagen's book, *Hitler's Willing Executioners: Ordinary Germans and the Holocaust,*[2] describes just this—the murderers in the concentration camp, as well as the accomplices who participated by looking the other way, were John and Mary Doe. They were faced with the choice to either speak up and be murdered themselves, or to do their best to keep it out of their minds, making poor attempts at justifying it, and basically pretending it was not happening. "Maybe they'll turn off the heat before the water comes to a boil," thinks the frog.

Such is the condition of the present state of the world—almost entirely void of contact, intimacy and aliveness. For those who have come face to face with this, it is a shattering contemplation.

> *The nuclear man is a man who has lost naive faith in the possibilities of technology and is painfully aware that the same powers that enable man to create new lifestyles carry the potential for self-destruction...He sees that in this nuclear age vast new industrial complexes enable men to produce in one hour that which he labored over for years in the past, but he also realizes that these same industries have disturbed the ecological balance and, through air and noise pollution, have contaminated his own milieu...He suffers from the inevitable knowledge that his time is a time in which it has become possible for man to destroy not only life but the possibility of rebirth, not only man but also of mankind, not only periods of existence but also history itself. For the nuclear man, the future has become an option.*[3]
>
> - Henry Nouwen

12 STEPS TO AVOID HUMAN TOUCH AND INTIMACY

In a roundabout fashion, we in the Western world actually practice how *not to touch,* not to feel, and not to care too much. The sum total of the following twelve principles are a description of contemporary life in Western culture. Granted, not all of them apply to everybody. And granted, most people would never consciously aspire to such a state, but if you examine these principles closely and honestly, you are likely to find that at least a few of them are guiding factors in your life.[4]

1. Eat, exercise and dress in a way that will make you feel and appear strong, hard and invulnerable. If you are a woman, starve yourself and wear a lot of make-up and fancy clothes. Live according to an image of who you think you should be and not as you are. (Plastic surgery is recommended.)
2. Find a job in which you will be overworked and stressed out. Find a mate who has a job in which he or she will be overworked and stressed out. Become too busy to have time for anyone or anything else. Become too busy even to spend time with yourself, so that you won't have to acknowledge your present state.
3. Pursue a profession in which it is considered normal and even desirable to manipulate, swindle and gain the upper hand in the matter, any matter. Enter a field that causes direct or indirect harm to others such as the alcohol or tobacco industry, oil production, the armed services, etc. All of this will help to desensitize you to the depth of human need and sensitivity.
4. Avoid exercise, fresh air, and nourishing foods so that your body will feel weak, toxic and unworthy of touch.
5. Spend as much time as possible in front of the television, computers and the like, filling your mind with images of murder, rape and unrealistic relationships so that you do not measure yourself by anything realistic or genuine.
6. Drink alcohol, smoke cigarettes, and take antidepressants and other socially acceptable or unacceptable intoxicants that will numb your mind and body, obscuring your ability to clearly perceive your present state.

7. Become part of a social circle in which people are pretentious, polite and/or superficial. Be sure to live in a culture in which, as a norm, people are generally unaffectionate with one another. Or, disengage yourself from social interaction entirely.
8. If you have children, be sure to place your own needs above their needs. Get babysitters every weekend. Buy them toy computers and allow them to watch as much television as they would like. This will help you to avoid the needs of your child, the depth of your responsibility to future generations and facing the innocence you have lost.
9. Buy a dog or a cat and place all of your love, affection and intimacy needs into that animal, while you avoid giving your loved ones this same attention.
10. Make a general attempt to deny your own feelings and your own pain. In doing so you will not notice the feelings of others, and if you do come across painful feelings in others by chance, you can simply blow it off to, "That's life."
11. Avoid close relationships with members of the same sex. They can see beneath your cover and do understand you, and therefore may demand that you be real with them.
12. In sexual relationships, be aware only of your own needs and do everything you can to get them met. (Isn't that what the psychotherapist said?) In doing so, you can avoid placing attention on your partner and his or her feelings and needs. Be self-centered and selfish in relationships in general.

Unfortunately, the above twelve principles describes us pretty well. Western society advertises a "look" that demands people to dress in a highly specific way in order to create a particular image, and to gear their eating and exercise habits (as well as face lifts and liposuction treatments for the rich), to create a body that matches it. Women have either starved themselves into skinniness or aspire to, and cover their faces with make-up (there are now "permanent" cosmetics available, thus women will never have to be seen "without their face on" again). Many people are stressed out by their jobs—their 40-hour work weeks coming nearer to fifty or sixty hours after they have commuted and put in overtime. Almost half of these workers are also women who have full-time responsibilities to their families. Many people have jobs which at best do not contribute to their society, and at worst stand directly in contrast to their moral integrity. Many do not have time to exercise, or do not live in places in which they are exposed to fresh air, and everybody inevitably takes in a tremendous amount of chemicals and pesticides from their diet—even those who primarily eat vegetables and fresh fruits. Close to 100 percent of Americans own television sets, and a growing majority own computers. Though cigarette smoking is generally decreasing in spite of the recent resurgence of smoking as a fashion, alcohol consumption is rampant and Valium™ and Prozac™ are taken without a second thought. People often find themselves in social situations that do not allow them to be vulnerable, expressive and genuine. Whereas some have the good fortune of one or two genuine friends, many do not. Because of growing financial demands, most parents have increasingly less time to be with their children, and a growing majority of children attend some form of daycare and are placed at the mercy of devices such as televisions and computers to serve as their "baby-sitters." Meanwhile, millions of dollars are spent yearly on the purchase of dogs and cats—these pets becoming the recipients of the attention and affection that children long for. (In the United States, more than one billion dollars is spent on pets every year.[5]) Men tend to have few, if any, close relationships with other men, and sex is used far more frequently as a commodity, a drug or a habit, than it is to invoke states of communion and to express genuine love.

Fortunately, there is more to life than these twelve bad habits—people still continue to fall in love, mothers hug their children and teach them to swim, attempts are made to live honestly and with integrity. But generally speaking, the trend is heading in one direction and you can guess what direction that is!

The question becomes, "Could it be otherwise?"

I say yes—that with a great deal of conscious attention, something else could be created. There could exist a culture of people who take good care of themselves—eating well, living naturally, exercising regularly, and maintaining healthy bodies while not being obsessed by their outer appearance. They would be people who worked hard, but in sane environments, and not so hard that they neglected their relationships with their husbands, wives and children. They would be people who indulged in the use of computers, television and toxic distractions with moderation, and who entertained themselves with one another's company. Women would find support and friendship with other women, and men through men. People would not consume excessive alcohol and drugs, as the richness of their lives would not dictate the same need or desire for such substances. There would exist a set of values that would encourage mothers and fathers to learn to really *be* with their young children—for mothers to breast-feed and for both parents to nourish them with lots of touch and affection. People might have pets in order to bring greater enjoyment into the family environment, but not to unconsciously supplement or replace their need for affection from one another. Time to reflect, relax or otherwise stay connected with oneself would be valued, and adults would aid one another in healing their own psychological wounds so that they would not pass them on to their children. Relationships would be based on giving instead of taking, and sex would be considered sacred. It would be a generally wholesome, life-giving, touch-giving, love-giving environment—an environment of sanity. Need this possibility be so far away from where we presently stand?

IMPERSONAL EVERYTHING AND THE COMPUTER AGE

Customers of a popular Chicago bank are currently charged three dollars each time they choose to do their banking interactions with a human bank teller instead of a machine.

Children can go to their bedroom after dinner and punch a few buttons on their computer and be met with screen after screen of full colored pornographic images.

Scientist Masuo Aizawa of Japan is growing colonies of nerve cells that he one day hopes will solve specific problems—he is designing "living brains."[6]

First the industrial revolution sped up society and reorganized the entire occupational structure of the industrialized world, taking many people out of work but also creating many new jobs. Now the computer revolution has come, and human effort is becoming increasingly less valued, while computers are already being considered as a "bare essential." One job I was hired for had a requirement that all employees have an answering machine for their home telephone, or they couldn't work there. Of course, I did have an answering machine—in fact, both myself and my two other housemates each had our own telephone line and answering machine...but I was struck by the fact that one had to have a $90 machine that was only popularized in the past ten years, in order to work in a counseling agency. Already in the late 1980s professors at many large universities were *requiring* all papers written by students to be printed out on a computer—typewritten papers would not do!

Computers are here to stay, there is no question about it. One could have endless intellectual debates as to the advantages and disadvantages of computers and still arrive back at the same initial dilemma, but it is worthwhile here to consider how we are being influenced by the increasing depersonalization of our lives, and what kind of effect this has on our basic need for human contact.

I do not claim to have the answer to this, nor am I a dropout of the age of technology. As an author, I am as dependent upon my computer for my livelihood as most computer programmers are. Computers are convenient. They save time, labor, effort. They make our lives easier to manage, and our "handwriting" easier to read! There are benefits to computers and all forms of modern technology, and I will not attempt to argue to the contrary. Nonetheless, the potential outcome of the direction that we are headed fills my mind with frightening images, and my gut with sadness and dis-ease. I sense that the consequences may be far greater than anything we are prepared to endure.[7]

When science begins to use computers to imitate human life—as is presently being done in the field of artificial life—we are clearly becoming more and more removed from our basic humanness. I was shocked to learn that scientists have funneled the tremendous capacity of computers to not only create robotic type creatures that can imitate human functions, but beyond this are combining protoplasm, cells and molecules in such a way that they function independently as their own life force... they are attempting to create life itself. In his article, *Playing God—*

The Making of Artificial Life, David Freedman warns, "What if humans become capable of slapping together new forms of life as easily as a kid makes toys out of Legos? What if real brains can be assembled to do a given job? What will we do with that kind of power?"[8] And what will we do when we have successfully found a way to substitute computers for the vast majority of the work force? What will people do then, and more importantly, how will they feel about this?

Such radical scientific experimentation isn't just happening in the laboratory. The caption under an article entitled, "Touch Goes High-Tech," printed in *Psychology Today,* reads:

> *Here comes teledildonics, where your wildest fantasy becomes virtual reality. But will the desire to jack in and jack off make people blind to everyday problems?*[9]

It is proposed that by putting on a lightweight spandex suit, goggles and gloves that are connected into a computer system, a user can enter an alternative "reality." Tactile, audio and visual sensations will be created in which the user (through his body suit and the computer screens inside his goggles) will be able see, touch and be touched by the sex object of his choice. It is likely that he will also be able to experiment with changing his gender, playing out S&M fantasies, and even raping or murdering his sexual partner. Sound fun? Proponents of teledildonics, only one of the many forms of "cybersex" that is expected to be available through the Internet, claim both that "the opportunities for gender exploration and...increased sexual tolerance are enormous," and that it is a form of safe sex (if it could be called that). Critics say that not only is it no more than "an advanced form of masturbation," but furthermore that "the many uses of virtual reality, and especially teledildonics, will bring with them a whole new world of technologically induced trauma and disorientation."[10] Previously, if a man wanted to buy sex he would just go to a whorehouse and get what he wanted. It may have been crude, but it was simple and real. It was skin to skin. Now people can have sex on the phone, or sex with a bunch of microchips and wires. And cybersex is not always sweet and innocent. After analyzing 917,410 sexual images on the computer, Marty Rimm, a researcher at Carnegie Mellon University, concluded, "I think there's almost no question that we're seeing an unprecedented availability and demand of material like sadomasochism, bestiality, vaginal and rectal fisting, eroticized urination...and pedophilia."[11] Is this the kind of touch we are headed for in the millennia?

Furthermore, there is an increase of cases such as that of the thirteen-year-old girl from Kentucky who was found in Los Angeles, having been lured there by a "cyberpal." According to Detective Bill Dworn, head of the Sexually Exploited Child Unit of the Los Angeles Police Department, "The pervert can get on any bulletin board and chat with kids all night long. He lies about his age and makes friends. As soon as he can get a telephone number or address..."[12]

The implications of these trends on an already touch-starved nation are eerie and frightening. It was one thing when people were going to newsstands and purchasing copies of *Playboy,* but now people of any age can get on-line and tap into any number of libraries of pornography, sexually oriented chat lines, and seemingly real tactile sexual encounters. Wounded individuals, whose bodies are screaming for physical contact in order to bring them back to their humanness, are being given yet another opportunity to escape into a tactile virtual fantasy, which in the greater scheme of their lives, they will pay for dearly through the increasing impersonalization of their lives.

An Increasingly Impersonal Society

In a deeply personal society, no one would be rewarded for doing their banking with a machine instead of a person, nor envied for having sex with a computer instead of a human being. In a deeply personal society, none of this would even be considered. But we do not live in a deeply personal society, and when we fail to touch one another, the society around us will reflect this by becoming equally impersonal.

I was not quite prepared for what the bank teller would tell me and two eight-year-old children on a recent field trip to the bank. We were studying money, and in response to one of the children questioning her about what a world without money would be like, the bank teller described to the children the concept of the debit card that will eventually make money obsolete. She went on to fill their innocent minds with the likely proposition that individuals will have small computer chips surgically set into their arms. The chips will be connected to their debit card and will serve many other functions such as that of an identification card, a key to one's home, and so forth. Needless to say, the children's conversation on the ride home was not about the origins of money, but about what it would be like for everybody to get computer chips implanted inside of their arms!

I was similarly shocked by a visit to the Arizona State University library where I went to do some research on the field of touch. I entered the library and found a woman at the information desk. When I told her what I was looking for she responded, "Oh, you should be able to find that in the sociophile on the silver platter." When I laughed in response, believing she was telling me some sort of techno joke, she gave me a blank look. She then explained that this was the name of a specific research program on a specific kind of computer. Now wary of the success I would have on this expedition, I walked through corridors of computers until I came to the proper set. I spent the next two hours learning how to use the program in order to access the desired materials, and soon discovered 2064 listings for "touch" in various libraries throughout the state. After spending the following three hours searching through about 30% of the titles, I was exhausted. Next, I endeavored to learn another program that would inform me *where* these articles could be located. Hours later, after going up and down elevators, in and out of rooms, purchasing copy cards, and talking with still more computers, I collapsed on the steps in front of the library with two short articles as my trophies for the day's work, pondering the growing complexity, impersonalization and lack of human contact even in public facilities such as libraries. I ended up doing the majority of my research at the Prescott Public Library, where I could walk from one end to the other without getting lost and where the reference librarians were willing to talk to me instead of insisting that I communicate with a computer!

Western medicine has become yet another impersonal institution. A friend of mine who is a medical doctor in India was shocked to find out about the relationship (or lack of relationship) between most doctors and their patients in the West. In India, a doctor has an investment in his or her patient as a person. He diagnoses not only the symptoms of the illness, but familiarizes himself with the person behind it, his or her life circumstances, and his or her individual needs. When he takes on a patient with an illness, it is his personal and ethical responsibility to see this person through to its cure. He will not continue to charge a flat rate of $95 per fifteen minute slot for each follow-up visit, particularly if he knows that the patient can not afford it. Even so, Indian doctors are well off financially. Institutionalized medicine in the West stands in direct contrast to this.

You can guess with a fair degree of accuracy what you will find when you go into a Western medicinal institution. The smell of sterilized gloves, antiseptics, deodorizers and sanitation chemicals fills the air. Here you will be greeted by a polite receptionist (polite does not equal personal), given a clipboard of forms to fill out and be expected to wait—often for extended periods of time—even if you have your three kids with you or an important meeting in the afternoon. Perhaps if you have not stayed in a hospital for an extended period, you do not know that patients are not spoken of by name, but are instead classified either by their illness or their room number. You may or may not be touched by your doctor or nurse, as the touch codes in hospitals are continually becoming stricter, and if you are touched it will probably be by a plastic surgical glove covering a hand. In the words of one physical therapist who works in a hospital setting, "We are so fragmented ourselves, so disconnected from our own bodies, that we don't hesitate to separate a person from his body part, and to treat him as a broken leg or a sprained wrist instead of as a human being."

And not only is our culture imprinting upon us in 1001 ways that touch is dangerous, that bodies are dirty and unsanitary...but we are buying the party line! We are so afraid of our own sweat, blood, germs and minor imperfections (as evidenced in an excessive use of antiperspirants and deodorants, unneeded anesthetics, scented sanitary pads, etc.) that we don't want to touch, or be touched by others. I was quite surprised when I walked into the bathroom of a popular family restaurant chain and found it "touch-free"—no doors to open, a toilet that was automatically programmed to flush as the individual left the stall, soap that drips on one's hands when they are placed under the dispenser, water that mysteriously knows when to turn on and off, and a hand-dryer that starts automatically when hands are placed in its vicinity. I could not disagree with my friend when she supported this system, as it was certainly more germ-free, but I had recently returned from an Asian country in which everybody touches everything and everybody, and the whole thing struck me as baffling—not only impersonal, but simply strange. Although the population is exploding at unprecedented rates, we seem to be living more and more in a society without *people* in it.

When people themselves begin to become impersonal, our lives become the casualties of the present state. This was glaringly obvious to me when I returned home from spending a year in Asia. As soon as I got on the airplane, the pilot began, "On behalf of Singapore Airlines, welcome!" The very pretty stewardesses brought me very neatly arranged, plastic-tasting food. I had an insignificant and superficial conversation with the man sitting on my right, and did not speak a word with the woman on the left for the fourteen-hour duration of the flight—each of us acting as if the other person was invisible. When we landed, America suddenly appeared as a horror of a "friendly" place. Everywhere I went—gas stations, supermarkets, and restaurants alike—everybody sounded like a parrot who had been trained to speak in "friendly-language."

I have walked in two worlds. Half of my adult life has been spent tromping around rain forests in knee-high boots, bathing at a well in the company of chickens and mosquitoes, lighting a candle so I could read before I went to bed, spending an hour just trying to get a letter mailed, and walking crowded streets with my eyes on the ground to avoid walking straight into animal shit! The other half has been spent in educational institutions, sitting thousands of hours in front of a computer, racing around in my car from appointment to appointment, trying to keep up with the rat race, playing the game of big business. And I must say, the *quality* of my relationships with people, and my own feelings of awe and wonder with life, have been directly proportional to the tacit humanness that existed in whatever culture I was living in at the time. Again, it is totally impractical for the vast majority of the people to go to Third World countries and walk around in bare feet with a machete in hand to try to get in touch with themselves, but let us not fool ourselves into thinking that we are living in a personal world.

I think that human beings were placed here together both because we have some business to do together, and because we need each other. We cannot get the same physical nourishment from spending a Saturday evening on the Internet looking at pictures of naked women as we can cuddling with our husband or wife, or by spending a quiet evening with close friends. It is simply not as satisfying to heat up a Lean Cuisine™ in the microwave after work as it is to have a home-cooked meal. It is not the same to drive into an automatic, credit card operated gas station without an attendant, as it is to drive up and pump our gas when somebody is washing windows, somebody is cleaning the cement, and somebody else is collecting the money. These small details of an impersonal world, in their sum total, take a significant toll on the overall quality of our lives.

Human beings thrive from the nourishment, comfort, encouragement and joy that they receive from their contact with others. They grow from being confronted with failures in communication, through acknowledging the difficulties in loving one another, and by getting direct and indirect feedback from their world and from those whom they come into relationship with. Almost all religions of the world agree that the purpose of human experience is to learn to love one another, to grow and to help others. The possibility for the fulfillment of this purpose is diminished when the entire day is spent in relationships that lack depth and do not acknowledge the essential nature of the people involved in them. People in the Western world have devoted massive amounts of creative energy, time and money into complex technological pursuits. How would it look if even a fraction of this energy was devoted to creating an environment that was full of touch, intimacy, affection and an overall livability that would provide nurturance and sanity to those who wish to sustain their humanness in a time of growing alienation and modernization?

THE UNTOUCHABLES

> *They know that the world does not much like them and they try hard to be good to one another.*[13]
>
> - Jonathan Kozol

The term *untouchability* commonly conjures mental images of poor, hungry, and dirty people in India who belong to the class of "untouchables." These people are relegated to the lowest jobs, if any jobs at all, such as toilet cleaning and street sweeping. They are considered as dirt by their society—as less than others; almost as flies or stray dogs that can be cast off by the flick of a hand. Higher castes in India consider it contaminating to have any physical contact with an untouchable.

Yet untouchability is not only alive and active in India and the East, but is operating in full swing in our own backyards. Who are the untouchables of the civilized world? They are the gay people, the African Americans, the Hispanics, the Chinese, the poor, the homeless, immigrants, people with AIDS, the elderly, people with physical and mental disabilities, and people who live in communities, spiritual groups and other alternative lifestyles. They are the "have-nots" in any given society. They are all those whom we try to cast off into the corners of society where they will not be seen, such as ghettos, or into positions in which they can not be heard, thus keeping them out of spheres of public and political influence where they would have a voice. They are the people whom we would rather not think about (because they make us uncomfortable) and certainly not talk about, for to do so inspires feelings of disease, guilt, fear and irreconcilability.

Terms such as racism, sexism and homophobia have become popular fronts for our widespread prejudice. "Politically correct" language has become incorporated into our vocabulary such that many people now use the term "woman" instead of "girl," "Afro-American" instead of "nigger," and "gay" instead of "fag," without even knowing why they are doing so. However, these names are often little more than superficial niceties that often serve to mask the underlying fact of socially sanctioned untouchability. Although there have been improvements in the last fifty years, racism is still breaking hearts and destroying lives every day; gay men and lesbian women are still being outcast by their families; the poor are waiting in the emergency room line for three and four days in order to simply see a doctor. Untouchability implies that there are entire factions of society that do not receive touch—both literal touch as well as the acknowledgment of being citizens worthy of all the privileges (e.g., welfare, housing, healthcare) extended to all other members of the culture.

The issue of untouchability is perhaps most poignantly illustrated in the AIDS crisis. Although it is most widespread among the gay population, AIDS has now extended into all corners of the modern world, as more and more movie stars, school teachers, friends, and other "ordinary" people die of this illness. Unfortunately, most people continue to link AIDS with homosexuality, and proceed to feel increasingly separate from, and prejudiced against, gay men and lesbian women. Furthermore, there are a great many widespread assumptions about AIDS that are incorrect, such as the belief that one can get AIDS simply by touching a door handle that a person with AIDS has touched, or by hugging someone who has AIDS. Oftentimes, even those who want to touch people with AIDS are blocked by their fears. A hospice worker tells the following account of her first contact with an AIDS patient:

> *It was my first hospice patient with AIDS. I was very uneasy, even though I had been trained and educated on the subject. I thought I'd get AIDS just by touching the pencil he had used. He had become demented because of the AIDS and would salivate and spit—I thought I might get AIDS from that. I never touched him. I knew that it was totally irrational. I had been given the facts, but I couldn't overcome the strength of my fears.*

In contrast to this stands Ma Jaya Bhagavati, a spiritual teacher whose Florida-based community runs a hospice for people with AIDS and regularly visits AIDS patients in hospitals. She tells the story of walking into a man's room and noting the immense sterility and de-personalized setting. The doctors who visited him were formal and distant, and the man was connected to a complex network of machinery. Using her connections to cut through some of the red tape, Ma Jaya Bhagavati was able to spend some time alone with this young man. She immediately climbed into bed beside him, and held and comforted him, giving him some of the only genuine human contact and affection he would receive in his last days.

Untouchability is a reality. Although we'd rather not admit to it, most people do not want to touch the hand of the homeless person, or hug the young mentally retarded child. As badly as we may feel about it, the majority of us do not want to spend time visiting with elderly people in a nursing home, or to serve food at the local soup kitchen. We do not want this skin-to-skin contact, and hope that somebody else will do it for us.

Do professional touchers exist? Are there people who are willing to do this work? There are a few. They are generally the underpaid workers of the government social welfare programs whose funding is constantly threatened. Most are so overworked, and so drained by their lack of ability to meet the needs of the people they are responsible for, that they often become burnt out, hardened and cynical. Many of these case workers are people whose own touch needs were never met, and who are knowingly or unknowingly trying to fill this void by "touching," i.e. helping others.

I am familiar with the life of the underpaid, overworked and burnt out social workers because I was one of them. I worked in an abortion clinic as a bilingual assistant to the medical doctors who performed abortion. I did blood tests and pregnancy tests, handed tools to the doctors, cleaned up after the abortions, and provided emotional support for the many women who were traumatized by the surgery.

I worked for less than seven dollars an hour at the hardest job I have ever done in my life when I could have been making forty dollars an hour, and many times I asked myself what was motivating me to do it. It is years later, and I now see that my own touch needs were in some way fulfilled by serving those who were otherwise not receiving healing touch. I did not become a professional toucher because I was altruistic; I *needed* to do so in response to my own feelings of touch-starvation.

Professional touchers sustain the hospices, the hunger programs, the nursing homes and the drug treatment centers for teenagers. Many times in these places you will find one or two individuals—usually doing dishes, cleaning the toilets, or patiently holding one of their clients as they cry—who have very consciously dedicated themselves to the task of caring for those whom society has rejected, knowing that while their assistance will not produce change on a large scale, that it is having a significant effect on the few people with whom they come into contact.

In India, you cannot walk down the street without being met by several "untouchables." How then, in contemporary industrialized nations, do we hide our untouchables away?

Author and social critic Jonathan Kozol reports standing on a highway overpass looking at the New York city ghettos and being confronted with murals showing interiors of pretty rooms, window shades, curtains and pots of pretty flowers—all painted on the sides of buildings that face the highway. These pictures were "done so well that when you look the first time, you imagine that you're seeing into people's homes—pleasant looking homes...that have a distinctly middle-class appearance." The social worker he was with commented that the city had these murals painted on the wall "not for the people in the neighborhood—because they're facing the wrong way—but for tourists and commuters...The idea is that they mustn't be upset by knowing too much about the population here. It isn't enough that these people are sequestered. It's also important that their presence be disguised or 'sweetened.'"[14]

We hide from our untouchables with the numbness we have created in ourselves. We hide from the untouchables in our false hopes that our governments are taking care of them, in our ideals of democracy and freedom, in our prayers that "God" will tend to them, and in our wish not to see. We hide them from ourselves in nursing homes and in housing units in neighborhoods we do not drive by. We pretend that they are not there. We are too afraid of how we would feel if we did not hide. In the words of Pastor Martha Overall of St. Ann's Church in the South Bronx:

> *...Looking into the eyes of a poor person is upsetting because normal people have a conscience. Touching the beggar's hand, meeting his gaze, makes a connection. It locks you in. It makes it hard to sleep, or hard to pray. If that happened, you might be profoundly changed...Writing a check to the Red Cross or some other charity can't do that. [What many charity organizations are] really telling us is, 'Do not open up your heart. Don't take a chance! Send a check to us and we will do the touching for you.'*[15]

Having thus far focused on our unwillingness to see, it is equally true that we are not a callous, mean and uncaring people. It hurts us when we are faced with the devastating facts of the housing conditions of immigrant farm workers, the astoundingly high statistics of people who are HIV positive, the slander perpetrated on individuals for practicing their freedom to choose their own religion or the gender of their sexual partner. Although there still remains an appalling number of individuals who will say things like, "The faggot had it coming to him," or "The nigger is too lazy to earn a living," this does not express the feeling of the majority. Many people *do* care, they simply feel helpless to express this. Reverend Overall continued:

> *It has to take extraordinary self-deceit for people who plant flowers on Park Avenue but pump their sewage into Harlem and transport their medical waste, and every other kind of waste that you can think of...to imagine that they have the moral standing to be judges of the people they have segregated and concealed. Only a very glazed and clever culture in which social blindness is accepted as a normal state of mind could possibly permit itself this luxury.*[16]

Instead of exposing ourselves to this type of discomfort; instead of touching the untouchables with our own hands, we drop off our old clothes to the Salvation Army, we give a dollar to the homeless person, we smile at the retarded child once in awhile, and we carry on, figuring that there is nothing we can do, while a part of us dies with them.

Internalized Racism

Unfortunately, society's sentence of untouchability seeps into the psyche of he or she who has been labeled as such. It is common for social psychologists to use terms such as "internalized racism" or "internalized homophobia" to describe this phenomenon. For example, a gay man who is confidently living a comfortable life with his gay partner and who speaks easily about his choice of lifestyle, may harbor a secret shame that he is dirty or morally incorrect. (After several months of therapy, my outspoken, lesbian client came into my office in tears begging me, "Please tell me there is nothing wrong with me for being gay.") A Mexican-American woman who had difficulties in school because of strong cultural differences may come to believe herself to be unintelligent and unworthy of being heard. The combination of these internalized beliefs (which are rarely conscious to the individual) and the feelings which he or she holds toward the society which has allowed this to take place, often result in rage and hostility.

The consequences? The victims become perpetrators. A young black man robs a store to get money for the drugs he uses to cope with his inner hell; a Russian immigrant never learns English as a result of having been called stupid, and thereby becomes another statistic on the social security waiting list; the homeless man yells at the wealthy woman who just walked by him pretending that she did not hear his request for change. Thus, our prejudices are affirmed.

The Hope

Mahatma Gandhi, the revolutionary Indian leader and perhaps the most admirable advocate of social change in the twentieth century, spent a lifetime working for the improvement of India's untouchable class, often visiting them and performing tasks with them such as cleaning latrines, which was unheard of for a man of Gandhi's social class. He encouraged people to change the term "untouchable" to *harijan,* which means "child of God."

A Christian couple in New England adopted over a dozen refugee children of the Vietnam war—many of them severely emotionally traumatized—and integrated them with their own children. They now operate their "House of

Hope" according to a combination of traditional Christian and Vietnamese Buddhist religious beliefs and practices.

An influential and popular spiritual leader in India recently allowed an intercaste marriage to be conducted in his temple, much to the horror of many of his followers.

Are we willing to follow their example in our individual lives? Are we willing to simply admit to ourselves that there are large groups of people that we are afraid to think about and would rather not touch? In acknowledging our fear and prejudice honestly lies the seeds of change. Untouchability is yet another symptom of a culture that has been touch-starved and in which people no longer know how to relate with one another compassionately.

A LOSS OF TRUST

Can you trust your mother?
Can you trust your chief?
Can you trust your brother
—the one that killed the thief?
Can you trust the scientist?
Can you trust a friend?
Can you trust the holy laws, reinterpreted by men?

- liars, gods and beggars[17]

In 1995, Robert Putnam, a professor of International Affairs at Harvard University, released an article that caught the attention of the U.S. government and became the subject of widespread response and social concern. The article, entitled *Bowling Alone*,[18] spoke of the decline of what Putnam called "social capital," which refers to an individual's network of social connections. He noted with statistical evidence that while individually centered activities (i.e., bowling alone, computer games, Virtual

Reality) are on the rise, participation in community-based activities such as bowling leagues, PTA meetings, religious groups, Boy Scouts, civil and fraternal organizations, and voting are sharply declining. He quoted a poll in which Americans were asked if they could trust most people. In 1960, fifty-eight percent of Americans said that they did, whereas in 1993, only thirty-seven percent responded in favor of trust. What will this look like in another fifty years? At the current rate, nobody will be able to trust anybody.

A touch-starved nation very quickly becomes a trust-starved nation. The fact that Putnam's article caught the attention of the United States government, resulting in a series of talk shows and major magazine articles in response to it, is evidence of a real concern among the people. Suddenly a Harvard professor is telling us that people are out of touch with one another.

My initial visit to Puerto Viejo, Costa Rica, revealed to me how much I had absorbed the general mistrust that is pervasive in Western society. Having arrived directly from the winter blizzards of Ann Arbor, Michigan, my first days among these coastal people were very disconcerting. These people were open...and trusting! Every time I would pass anybody on the sandy roads, they'd look up at me with a bright smile and say, "W'appin man?" and often stop for a few words. Automatically my mind would begin to race, "What do they want from me?" "Why are they talking to me?" Yet this is simply how they greeted everyone they met, for it was a small village, everybody knew everybody, and they had apparently made the assumption that any stranger who cared to travel such a great distance to visit their village had to be all right.

By the time I arrived back in Ann Arbor several months later, I had been converted to the relaxed, trusting way of life I had found in Costa Rica. Naively, I imagined that I could act in this new way back in my old environment. As you can probably imagine, it didn't go over too well. Of course I wasn't saying, "W'appin man," to everybody I passed, but I was saying hello to strangers and with very few exceptions it appeared as though I was invisible to them, more often than not being met by the top of a head that was turned to the ground, as mistrustful of me as I had been of the natives. Soon enough, I was back to doing the same sidewalk stare as those around me.

The design of our present world is "dog eats dog"—eat or be eaten. Society has been set up in a way that is aggressive and competitive. People are constantly on the defensive, consciously and unconsciously hurting one another in an infinite number of small ways. In spite of an nationwide air of confidence and arrogance, people are extremely suspicious of one another, and given the circumstances, with good reason. But what are the costs of this? On the most basic level, we now see children on leashes that were until recently used strictly for pets—so that if they run into the street or a stranger tries to talk to them we can just tug on the leash and the child will come back. We find women walking around with cans of mace in their purses in order to fend off potential rapists. There are guns and knives in the bedside drawers of many, and not only to protect against criminals who live outside of the house, but inside as well. If somebody taps my shoulder from behind me on a busy street corner to ask directions, I am likely to jump in fear. Many youngsters walk around with spikes on their leather jackets, chains around their arms, and piercings in their eyebrows, lips and elsewhere. Through their appearance they are screaming out, *"Stay away, I don't trust anybody."*

The more we cease to trust one another, the more violent we are likely to become. Throughout this book, you will read about the consequences of the failure to receive the necessary touch from birth onward. The result is that we simply don't trust ourselves. Most people have never trusted themselves, and regardless of where it originated, this climate of terror is now rooted *inside* of us. We are not unaware of how often we feel rageful toward our husbands and wives, our pets, and toward other drivers on the highway. We secretly note the small ways that we are scandalous and dishonest as we move through our days. Believing that we ourselves are untrustworthy, we do not bother to act in a manner that is worthy of trust, and we lose touch with what is most real in ourselves.

Trust is going to have to start at home. When somebody is expecting the worst of us, that is often what they get. On the other hand, when we know that somebody trusts us...when they show us by the way they relate to us that they are respecting our integrity, our maturity, and our ability to make decisions and to act wisely, we in turn begin to act trustworthy. Similarly, when we give someone else our trust and respect, the chances of that person acting in integrity with us increase dramatically.

There is no easy answer to the present predicament of distrust. It is not as though we can simply begin to trust everybody we meet—it is both impossible as a result of our conditioning, and impractical in terms of the state of the world. However, we can become aware of the profound lack of trust within ourselves and acknowledge the effects of this on a touch-starved nation.

THE FAST LANE TO NOWHERE

> *Once poets resounded over the battlefield; what voice*
> *can outshoot the rattle of this metallic age*
> *that is struggling on toward its careening future?*[19]
>
> – Rainer Maria Rilke

So what do we have to show for our lives as we pull up to death's gate driving seventy-five miles per hour in our Ford pick-up; having finally saved enough money to take our grandchildren to Disney World; having worked forty hours a week for forty-five years at the factory; having taken in two thousand pounds of hamburger, Swiss cheese and donuts; having failed to pay off our mortgage but receiving a small social security check; having smoked for thirty years without getting cancer; having watched our kids grow up and become just like ourselves; having watched "progress" sail by us, leaving us in its wake wondering what has happened; having *gotten by...?*

Is this the ticket to the heaven we were taught about as children—where the white angels open the gate and take us to a cloud-covered paradise? (What kind of software does God use to keep track of everybody anyway?) Or, as others believe, do we just explode into blank nothingness, and so what if the world is going to pieces and my plastic milk carton will be around for three generations and my great, great grandchild will feel isolated and depressed in her radiated world—unbonded herself because her mother, and mother's mother, did not know that breast-feeding and continual touch and affection is an absolute necessity for raising a child who feels secure in the world? Is our touch-free, germ-free, computerized, God-free environment taking us to where we want to go? And more importantly, is it taking us to where our children's children want to go?

> *We wonder if we might not be making a very big mistake, missing that opportunity we have been waiting for all our lives. And even as we observe these things about ourselves, we feel the urge to defend them, to reassure ourselves that maybe it isn't as bad as all that.*[20]

There is a contract with fine print attached to the Western dream of wealth and prosperity. Even if we don't know it, we have signed on the dotted line. We have signed our lives over to an ideal that will not only never provide us with what we were really searching for, but one that promises to slowly consume our vitality and aliveness, just as the rising temperature of the water in the pot slowly takes the life from the frog.

Montagu says:

> *...The individual is gradually converted into a device with a built-in design for achievement in accordance with the prevailing requirements, entailing the suppression of emotion, the denial of love and friendship, the ability to trade with whatever serves one for a conscience, while conveying an unvarying appearance of rectitude.*[21]

Heaven or hell happens right here on this plane. In spite of the popular commentary on reincarnation, as far as most of us are concerned, this one life, and the conscious will to make of it what we can, is what we are given. And if we want to know how we are doing, all we have to do is look around. Joseph Chilton Pearce explains:

> *History doesn't necessarily repeat itself. Phenomena have appeared over the past fifty years that have no historical precedence, for which our genetic system can't compensate, and that have so altered our mental makeup that we are blinded to the obvious relation between cause and effect.*[22]

In spite of the greatest technological advances, we're heading down, and if a big hole in the ozone layer of the sky over Australia doesn't speak to that; if children idolizing cartoon characters that murder and destroy others doesn't speak to that; if the question, "Isn't there something more to life?" doesn't speak to that—nothing will.

The future of a nation that is either touch-starved or full of affection and aliveness, depends on one thing—our children. The child who experiences himself in his rightful place in the family, community and culture, will become an adult who will give more than he takes, and who will bestow affection freely. Feeling cared for, he *will* care for others. Touch begins right at the start.

CHAPTER 2

First Touch: Birth and Childhood

Touching, yes, is the root...
We have to feed babies,
fill them both
inside and outside.
We must speak to their skins...
which thirst and hunger
and cry
as much as their bellies.
We must gorge them
with warmth and caresses
just as we do with milk.[1]

- Frederick Leboyer

Marasmus, stemming from the Greek root meaning to waste away, is a condition which results in physical death in infants when they are not given sufficient touch. Nearly fifty years ago, Dr. René Spitz discovered that more than thirty percent of infants raised in orphanages in which they did not receive contact, care and affection from a primary caretaker did not survive their first year of life, in spite of adequate food, materially hygienic surroundings and excellent medical care.[2]

On the most basic, instinctual level, physical contact is essential to sustain all human life—a type of food that is as necessary to the infant's well-being as is the physical food he takes in. On a deeper level, the intimacy that is created through touch is what creates a feeling of "aliveness" in the individual—for touch brings us to life.

The contact that transpires in the first moments, hours and months of a child's life will be largely responsible for the difference between a human life characterized by a persistent undercurrent of lack and desperation, and one that is based on a sense of security and belonging. Essentially, first touch—not only the first moment of physical touch, but the touch that the infant receives at the beginning of his or her life—is the single most influential factor that will determine a future life of love, or one of unlove. Although it is true that children are unlikely to remember their births and early childhoods in later life, the impressions they receive and how their parents treat them in their formative years is literally who they will become in later life. First touch is *that* important.

Parents have an utter responsibility to the life they have created. In one way, the child is separate and independent from her parents—existing as her own entity. But, at the same time, the child is very much interconnected with her parents, and is wholly affected by the attitudes and behaviors that they enact, both toward her as well as toward each other in her proximity. All a child's relationships, for the rest of her life, as well as her good and bad habits, her attitudes toward others, her sense of belonging or alienation, emptiness or fullness, are given by the parent, and will be either the child's sentence or blessing to live out. (Of course a person can always choose a path of self-transformation in adulthood, but in unloved and unbonded children this consideration often never even arises.) If parents fully understood the implications of their influence on their child, especially in the beginning of his or her life, the necessity for abundant touch, nurturance, and affection, would never even need to be considered.

> *A baby deprived of the experience necessary to give him the basis for full flowering of his innate potential will perhaps never know a moment of the unconditional rightness that has been natural to his kind for 99.99 percent of its history. Deprivation, in the degree to which he has suffered its discomfort and limitations in infancy, will be maintained indiscriminately as part of his development..."*[3]
>
> - Jean Liedloff

This chapter emphasizes the relationship between the mother and child. In my own case, after years of trying to prove myself through accomplishing man's feats—both large and small, and fighting for a woman's right to, in basic terms, act like a man—I have come to respect that there are objective differences between men and women—necessary and wonderful differences. One such difference is what a woman is able to offer her child in the first period of the child's life, and what a man can offer. What an infant needs most in the womb, in the birth chamber, in those first hours, weeks, months and initial years, is her mother. However, this in no way excludes the importance of the father, which will be discussed in detail later in the chapter.

KNOWLEDGE LIES IN WOMEN'S BODIES

> *Who shows a child as he really is? Who sets him in his constellation and puts the measuring-rod of distance in his hand?*[4]
>
> - Rainer Maria Rilke

To the fault of no individual (yet simultaneously a calculated aspect of the consumer society's plan to sustain itself), women are systematically drained of their knowledge and intelligence. These powers, if allowed to express themselves freely, would provide the woman with all she needs to know to raise a child who, in turn, would be able to manifest his or her full capacity as an adult, uninhibited by feelings of shame and low self-esteem, and free from seeking other's validation of his or her basic "all-rightness." By instinct, such an "empowered" woman would raise a child whose natural state was happiness, health, intelligence and equanimity, instead of mediocrity, suppressed depression and a desire for power over others.

The knowledge of how to touch their infants is part of every women's knowing—programmed into her very cells, stimulated by the process of pregnancy and birth, and always available. The only reason that this knowledge is unknown to so many women is that contemporary Western culture—with its emphasis on a highly developed intellect at the cost of the knowledge of body and heart—has placed a greater value on the mind of modern medicine than on the mind of the body. Concerning the process of birth and early intimacy, the past century has shown a 180-degree turn away from the knowledge inherent in a woman's body, and an equally strong movement toward a technological method of producing children.

In the 1920s, world renowned behaviorist John B. Watson wrote his classic book, *Psychological Care of Infant and Child*—a book that was to "revolutionize" child raising practices for decades to come. In an effort to bring up a child "as free as possible of sensitivities to people," he suggested the following:

> *Treat them as though they were young adults...Never hug and kiss them, never let them sit on your lap. If you must, kiss them once on the forehead when they say good night. Shake hands with them in the morning. Give them a pat on the head if they have made an extraordinarily good job of a difficult task. Try it out. In a week's time...you will be utterly ashamed of the mawkish, sentimental way you have been handling it.*[5]
>
> - John B. Watson

Popular misconceptions about child raising support the idea that if parents satisfy all of the infant's needs, they are somehow indulging the child and setting up a system of dependency that the child will be locked into forever more. As a result, many mothers have lost access to their instinctual knowledge that would *never* allow them to treat their child this way, or to follow advice such as Watson's, to the letter. However, contrary to this mistaken belief, when the child's dependency needs are adequately tended to (not only the physical needs but emotional and psychological needs as well), they fall away naturally as the child matures and becomes capable of taking care of his own needs, paradoxically resulting in a *more* independent child. Montagu has noted that, "Anthropologists, looking at cultures where a baby's comfort needs are quickly and tenderly indulged—even before it cries—have found that by the third month of life these babies cry less than Western infants."[6]

The point here is neither a feminist condemnation of traditional child raising practices, nor a glorification of women's power. The issue here is touch—and the need to do whatever it takes to provide ample touch and affection in infancy and childhood so that we can produce a generation of adults who will have a clear enough sense of themselves that they will be able to create a more nourishing and sustainable environment for *their* children.

BONDING

> *Given a choice of life companions, I would take any number of brain-damaged people, rather than one unbonded person, for we are flexible beyond measure and can compensate for extensive physical damage, but lack of bonding finds no compensation.*[7]
>
> - Joseph Chilton Pearce

In order to fully grasp the lifelong significance of touch on the child's life it is essential to understand the process of bonding, for it is the basis of all the birth and child raising practices that will be discussed throughout the chapter.

A contemporary thinker who deeply understands bonding, and articulates it clearly, is popular author and lecturer Joseph Chilton Pearce. Of this he says:

> *Bonding gives an intuitive, extrasensory kind of relationship between mother and child. Bonding is a felt process, not available to discursive thought, language, or intellect...*[8]

And:

> *Bonding occurs when the infant is met on both physical and subtle levels by his caretaker. Anchored in the power of the subtle heart system, the infant is always anchored in the core of his life. He is rooted within the great subtle intuitive energies that power physical life, no matter how his physical situation shifts and changes.*[9]

Bonding is the most profound measure of touch. It is a fusing of hearts, a tacit connection to somebody outside of oneself, an intermingling of life-forces that allows communion to occur far beneath the level of conscious awareness and external circumstances. It is a process that occurs *between* mother and child, by which two separate entities become linked with one another and operate as one unit.

One father explained:

> *The whole term "bonding" really means passing along caring—a sense of love, where affection is obvious, where the child just feels it. The way you do that is in every single interface with them, no matter how they show up. You are love—not hippie love, not lustful love, but simple care and compassion.*

Mamatoto is a Swahili term for "motherbaby."[10] In contrast to the idea of mother and infant as two separate, but connected, individuals, *mamatoto* suggests that mother and child are operating as one. Though the task of fully satisfying the infant's needs is a demanding one, at times seemingly impossible, when the mother-child organism is considered as one, there is an absence of the tension that would otherwise be present if the mother thinks, "I am serving *you*," "I am doing so much for *you*," "I am providing for *you*." When mother and child are as one, what needs to be done is done.

Parents can educate themselves, but what occurs on a deeper level in the bonding process cannot be learned in any book. Bonding is physiological, psychological, emotional and intuitive. Nature has created a perfected system that is intended to insure the optimal maturation of the human being on all levels, thereby revealing a possibility for the experience of communion, selflessness and exchange of self with other through the process of childbearing.

The Mother Bonds Too

> *It is clear that the mother needs her baby immediately after birth quite as much as the baby needs her. Each is primed to develop their own potentialities—the maternal role in the one case, that of developing human in the other.*[11] - Ashley Montagu

Bonding is not a one-way street. The mother must bond with the child so that the child will be able to bond with the mother. The demands of mothering are going to be extensive. She will need to hold her baby when she is all "held-out," to respond gently and quietly when she feels like screaming. Though much of mothering is joyful, her patience is going to be stretched to the limit. What gives her the ability to extend herself to her child in the way that is being considered here is the strength of her connection to him and her love for him. Depth of bonding will yield greater ease and less hesitation in her movement to fulfill her child's demands and wishes. It will not be easy, but she will be willing to act because her connection to her child is profound and she knows how much her child needs her.

Therefore, the bonding process that occurs creates a sense of security not only in the infant but in the mother as well. When she experiences milk coming from her own breast—milk that she knows is sustaining her infant—she feels herself a woman. When she sees that her presence and touch can turn a wailing child into a calm one, she knows that her touch is power. The more she gives of herself to her child, the more she allows bonding to take over. Thus she becomes naturally devoted. Many mothers describe this process of bonding as a "rite of passage," discovering depths of womanhood within themselves that they never knew before becoming mothers.

The Unbonded

> *We are never conscious of being bonded; we are conscious only of our acute disease when we are not bonded or when we are bonded to compulsion and material things.*[12] - Joseph Chilton Pearce

Although the term "bonding" is frequently thrown around amongst medical practitioners, psychologists and parents alike, we as a culture do not understand the implications of it.

The reason that bonding is not understood is because so few adults were ever given a sense of real belonging, of knowing in the cells of their bodies that they were loved. (Just because a person can say, "Of course my parents love me, I know they do," does not mean that person has bonded.) If bonding is not an experience we know in the core of our being—if it has no more meaning to us than any other intellectual concept—how can we understand the full value of it, thus giving it the absolute priority that it demands? Instead, we understand the importance of bonding primarily through our own sense of unbondedness within ourselves.

A nurse-midwife told me the following story about an unbonded infant:

The mother was young. She had been through a lot of depression and seemed to want the baby for her own self-centered reasons. Her perception of a baby was how most of us would think of a puppy or a toy, and as an attention-getting device. Her interaction with the baby came very late in the pregnancy and was largely disconnected. Labor was very traumatic for her—she had a borderline personality disorder and her husband didn't have the skills to deal with it.

She insisted she wanted to breast-feed, but every time the baby would get hungry, the mother would go to the bathroom and stay there. Eventually she'd come out and try to nurse the baby, but I had never seen somebody try to nurse a baby without touching it. She wanted me to position the baby at her breast and hold it there. She would either put her arms up over her head or hold them by her side while I held the baby. All the nurses developed a way of positioning pillows so she could nurse the baby without touching it.

Fortunately, her grandma Edna came to the hospital the first two days. She rocked the baby a lot and stayed with it. When the baby went home with the mother, he was very uncomfortable. He would cry and act very colicky any time the mother was near him, but when the father or grandma was around, he would respond like a normal, happy, healthy baby.

The parents are doing pretty poorly and will probably end up getting a divorce. The father is afraid the mother is going to harm the baby. When she comes into my office carrying the baby, which she will only do when people are around because it brings attention to her—she will cross her arms to create a platform, clench her fists, and hold the baby away from her body. That's a baby who is unbonded and at high risk for abuse, but since she's breast-feeding, it makes a strong case for the mother to keep the baby.

If we as a culture genuinely understood bonding, there is *no way,* except in dire circumstances, that mothers would be given the amounts of anesthesia that are commonly administered during birth. It would be impossible to place babies anywhere but on the mother's breast in the first moments and hours of life. The conscience of our medical professionals would not allow babies to be placed in incubators except when absolutely necessary, and in these instances would only do so in the direct proximity of the mother. Bottle-feeding would be considered an option only when the mother could not breast-feed.

If we understood in our gut what was at stake for our child concerning the unfolding of his future life, mothers would take a stand in hospitals and insist on holding their infants immediately after birth. If we truly respected the necessity of bonding, we would give our children absolutely everything—not in terms of toys and clothes and birthday parties, but in terms of affection and attention, knowing how desperately they need this.

A child needs love. Love is not a thing. Love is life. It means being alive and caring for life. Life creates

> *needs. To fulfill them is to fulfill love. Love sustains life. All a child wants is to be loved.*[13]
>
> - J. Konrad Stettbacher

Our collective lack of understanding of bonding shows up in our radical misinterpretation of the infant's behavior. For example, when the doctor holds the baby upside down to make it breathe and it starts wailing, a parent may commonly think, "Thank God my baby is breathing well and that everything is fine. *I* am so happy." Perhaps that interpretation is correct, but perhaps from the perspective of the infant, everything is not fine. Perhaps the baby cries because he or she is terrified and in pain. Other times, parents may find themselves feeling resentful of the infant, "Why does she demand to nurse twenty times a day? Why is she so needy?" Or the parent may feel frustrated that his or her baby cries for an hour every time they put him down to sleep in a crib, drawing the conclusion that they have a "fussy baby." (Yet it is quite likely that the bonded infant will not wish to leave the body of his or her mother for the first two years of his or her life.)

The unbonded child will adjust to her unbonded state by a type of psychic death— becoming numb and indifferent even to her sense of loss and disconnection. However, the child will pay the price for her unbondedness in ten thousand ways throughout her life.

> *...For a child whose attachment needs are chronically frustrated, anxiety and depression will diminish when attention to the needs diminishes. The cost, however, is exorbitant: loss of the ability to trust, loss of the capacity for intimacy, and a diminished ability to empathize with oneself and others."*[14]
>
> - Paul Klein

The legacy of alienation and disconnection in the unbonded child is perpetuated by an ongoing protective mechanism. In order to survive the fact of not being allowed to bond, a person numbs himself to these early needs, becoming increasingly callous to both his own needs and the needs of others. For example, the unbonded mother is oblivious to, and perhaps even angered by, her infant's cries, not understanding them to be a direct message to her about his genuine needs. In other words, the individual's unconscious anger and resentment at her own lack of bonding continues to subtly pervade her experience as adults, leaving as a remnant an attitude toward their own children of, "I did fine without it, so can you."

> *The natural process of bonding between mother and child is interrupted in a way that is agonizing for the child and that would be for the mother as well if we had not been anesthetized by the unbearable level of suffering and isolation in our own childhoods and programmed—brainwashed actually—by a totally child-negative narcissistic culture at large.*[15]

If these attitudes were not unconscious, we would consider them to be cruel. But they are, and the best we can do is acknowledge the fact that whatever degree of bonding that we, as members of a touch-deprived society, were given as infants was not good enough, and that we have thus far been unable to provide better. It is only through this acknowledgment that we can begin to look at our present systems for bringing children into this world, and ask ourselves what aspects of these systems are responsible for sustaining this broken state. In doing so, we begin to take charge of creating something different in our lives.

BONDING IS NOT JUST SWEET SENTIMENT, IT IS A MORAL-SCIENTIFIC IMPERATIVE

> *The easiest and quickest way to induce depression and alienation in an infant or child is not to touch it, hold it, or carry it on your body. Extensive scientific research has documented that this form of sensory stimulation (touch and movement) during the formative periods of brain development is absolutely essential for normal growth and development of the brain and behavior.*[16]
> – James Prescott

Although the need for bonding is easily understood on an intuitive, emotional level, bonding is, in fact, a psychophysiological process. Although the bonding process appears to the naked eye as mother-holding-breast-feeding-infant-with-lots-of-eye-contact-and-tender-caressing, these behaviors are a mirror for underlying sensory-psychophysiological processes that are essential for the normal growth and development of the brain. This sensory stimulation is provided through nearness and intimate body contact between mother and infant. Joseph Chilton Pearce has said, "The issue is not sweet sentiment. The issue is intelligence, the brain's ability to process sensory information, organize muscular responses, and interact with the environment."[17]

Perhaps nobody understands the scientific implications of bonding more clearly than developmental neuropsychologist Dr. James Prescott, who has devoted more than thirty years of his life to scientific research and writing about the necessity of tactile stimulation (touch) and body movement stimulation in the development of the functional, non-violent human being. (For more of Prescott's findings, see section entitled, "The Roots of Violence" in Chapter Four.)

Prescott's initiated brain studies were inspired by the research findings of the Harlows and their associates that began in the 1950s and which documented the abnormal behaviors of maternally-deprived monkeys. In one experiment, described by Ashley Montagu, five female monkeys, of an ordinarily affectionate species, were raised without their mothers. When these test monkeys matured and gave birth to their own infants, three of them turned out to be highly indifferent and unaffectionate as mothers, almost totally unresponsive to their infants, and the other two were brutal and abusive to their infants.[18]

His own extensive research on isolation-reared monkeys (those raised in solitary cages with no physical body contact between mother and other monkeys) led Prescott to rename the "Maternal-Social Deprivation Syndrome" the "Somato-Sensory Affectional Deprivation" (S-SAD), which refers specifically to infants who are raised without sufficient body contact stimulation with their mothers. He explains that SAD results in the abnormal development and function of the brain. "During formative periods of brain growth, certain kinds of sensory deprivation—such as lack of touching and rocking by the mother—result in incomplete or damaged development of the neuronal systems that mediate affection (for instance, a loss of the nerve-cell branches called dendrites)."[19]

The cerebellum is widely understood to be responsible for the regulation of motor functions of the body including the maintaining of body balance. Not so well known is the role of the cerebellum in the regulation of the autonomic nervous system and emotional-social behaviors. In fact, the cerebellum regulates both painful and pleasurable sensations and states. Infants who are not rocked and cuddled are at risk for incomplete development of the brain pathways and systems that mediate pleasure (limbic-frontal cortical-cerebellar complex). Adults who have difficulty experiencing pleasurable states are more easily frustrated, more prone to depression and violence, and more vulnerable to drug or alcohol abuse and addiction.[20]

Prescott noted that monkeys raised in social isolation without physical body contact with their mothers or other monkeys would experience abnormal development and functioning of those brain structures associated with emotional-social and sexual behaviors. A number of brain studies conducted by other scientists on mother-deprived adult monkeys who were pathologically violent, confirmed the existence of these brain abnormalities. Similar behavioral abnormalities have been seen in hospitalized children as well as those raised in institutions. In other words, "When we don't pick up our kids, this results in the incomplete development and functioning of the cerebellum and related brain structures."[21]

The developing brain—like the body—is very immature at birth and needs sensory stimulation for normal growth and development. For the cerebellum, this means continuous gentle movement stimulation. Therefore, handling the baby—rocking and carrying it—is necessary for its brain to develop properly. Babies who are raised in institutions that are overcrowded and understaffed often do not develop the same intellectual capacities as other children do because the caretakers in these places tend neither to touch them as much as their parents would, nor have the same personal and emotional investment in them.

Joseph Chilton Pearce has subsequently affirmed these basic principles from a different perspective:

> *This limbic structure must be activated during the bonding period at birth. Unless it is activated, the relationship between the heart and the midbrain*

> *area breaks down. From then on, this heart-mind relationship is compensating rather than functioning, and there is a serious dysfunction in the way the person can relate to others.*[22]

There is a substantial body of scientific data on how the early physical environment influences brain development, function and behavior. It has only been touched upon here to highlight the facts of the psychophysiological foundation in the bonding process, and to emphasize from a scientific perspective the necessity of body touch and body movement in the mother-infant relationship in order for bonding to take place.

The Politics of Bonding

In addition to his research, the story of James Prescott's career is an interesting one, in and of itself, revealing the politics of a touch-starved nation. For nearly fifteen years, Prescott worked at the National Institute of Child Health and Human Development (a branch of the NIH) as Health Scientist Administrator, performing the research cited in the previous section which clearly suggested that the origins of human violence stem from the failure of the mother-infant bond. During this time, along with receiving numerous awards, he was invited to speak to the U.S. Congress and to the Senate of Canada, and had hoped to inspire the passage of new legislative bills. However, in 1980 he received unexpected word from his superiors asking him to step down from his position—in other words, he was fired. It was said that he was misappropriating the allotted funds for research into areas that were "not within the mission of the NIHCD." In subsequent years, documentation of his research has "disappeared" altogether from the organization's published findings.

It is my opinion that Prescott's studies were providing definitive evidence about the necessity for mother-infant contact during the bonding period which, if clearly understood and accepted by Congress and made public through the media, would necessitate a significant change in presently existing U.S. government systems ranging from welfare, to education, to women's health policies, to social services. Of course, this would require an increase of government funding in the exact places where the Federal Government is clearly aiming to cut back—a distinct threat to our present system. Furthermore, what those who were involved in removing Prescott from his post were probably not aware of was that they, too, were personally and psychologically threatened by the reality of their own early lives. No properly bonded individual could take such a stand towards the suppression of such badly needed information. Once again, it is not the fault of any one individual—what happened in Prescott's case is merely one more of the many symptoms of a touch-starved nation.

HOW TO ENCOURAGE BONDING

Having understood the sheer necessity of bonding, the question then becomes, "How can a mother best ensure the likelihood of bonding to occur, thus fulfilling the touch requirements for her child?"

The answer is very practical. For although bonding cannot be measured on a scale or sold in a package, there are specific things a parent can do. It begins in the womb, continues into the birth chamber, and is sustained by what I will refer to later in the chapter as "The Three B's"—breastfeeding, backin' da baby, and bedtime.

In the Womb

The field of fetal awareness began to receive popular attention in the seventies. In addition to his own experiments and extensive experience, Dr. Thomas Verny took it upon himself to devote six years of life and travel to meet with the experts in this field across the globe. In his comprehensive book, *The Secret Life of the Unborn Child,*[23] he summarized the evidence of hundreds of scientific studies with the following facts:

- By the sixth month of life in utero, and perhaps even earlier, the child leads an active emotional life.
- The mature fetus can see, hear, experience, taste, and even learn in utero.
- What a child feels and perceives in utero begins to shape the attitudes and expectations of himself that he will maintain throughout his adult life. The main source of these messages is his mother—not necessarily every thought or feeling she has—but her general mood and feelings about her child.
- The father's feelings about his wife and the unborn child are an important influence on the unborn fetus.

Verny further quoted a study done by neuroscientist Dominick Purpura of New York's Albert Einstein Medical College that revealed that the child in the womb has already formed the brain structures necessary for learning, and even awareness, between the twenty-eighth and thirty-second weeks of development.[24]

Bonding begins in the womb. When it is understood and accepted that the infant is "conscious" in the womb—perhaps not alert and aware in the same way as adults are, but nonetheless conscious—the need for maternal love, touch and bonding, becomes increasingly clear. The physiological functioning of the womb is designed to respect the "consciousness" of the unborn child. If not impeded by a mother who abuses her unborn child in the womb by smoking, alcohol or drug abuse, by living in toxic and abusive environments, or by holding a deep resentment toward the unborn child—the womb itself will nurture the unborn fetus, encasing the child in the natural intelligence of the mother's body. On the flip side—the modern world into which the baby will be born *will not* respect the child's consciousness. In fact, it won't even acknowledge it. Contemporary culture (which its members unknowingly subscribe to) holds a view of children as less than adults, as unintelligent, and thereby proceeds to mold the child into a model of itself—a model based on greed, consumption, fear and apathy. (See section entitled, *A Culture of Abuse*, Chapter Four).

Among the hundreds of extraordinary accounts Verny has documented to illustrate the consciousness of the child, is the story of Boris Brott, former conductor of the Hamilton Philharmonic Symphony. Even as a young man, Brott had been mystified by his uncanny ability to play certain pieces of music that he had never seen before. While conducting a score for the first time, he was often able to know the flow of the piece before he even turned the page. In particular, he would know the precise line of the cello. At a certain point, he mentioned this to his mother who was a professional cellist. When he told her the specific scores that he knew, she was not surprised, confirming that those were the precise scores she had played while pregnant with Boris.[25]

In one study, Dr. Michael Lieberman showed that an unborn child will become agitated (as measured by rapid heartbeat) each time his mother even *thinks* of having a cigarette—she needn't even put it to her mouth. He suggested that while the fetus can not intellectually comprehend that his mother is smoking, he is sophisticated enough to react to the sensation that will be produced in him due to the drop in his oxygen supply from the blood that she will provide for him through the placenta.[26]

The question therefore becomes, "How, in pragmatic terms, can the mother be with her child in a way that respects the child's already conscious state?"

Beginning with the obvious, the mother refrains from indulging unhealthy eating habits and from the intake of any type of intoxicant. She receives prenatal care and follows the doctor's or midwife's recommendations. She exercises and practices breathing and relaxation.

In addition, the mother comes to appreciate that she is connected with her child not only through the obvious intimacy that exists between them, as well as the food that she provides for them both, but also through their shared hormones and the physiological changes that emotions produce in the body. Dr. William Sears, one of the forefathers of conscious child raising, explains:

> *Mother and baby are part of the same hormonal network. Those same hormones which produce stress reactions in the mother (increased heart rate, increased blood pressure, flushing, sweating, headaches, etc.) also pass through the placenta to the baby. So when mother feels upset, baby is upset. Researchers theorize that if an unborn baby is continually exposed to stress hormones from the mother and is continually producing his own stress hormones in response to an anxious environment, he has a higher risk of developing an overcharged nervous system.*[27]

From this example, it becomes apparent that depressed mothers most often give birth to depressed babies, anxious mothers give birth to anxious babies, happy mothers to happy babies, etc. If the mother does not want her baby, the baby will, on some level, know it.

What, then, does a mother do if she finds herself in a circumstance in which she feels unable to properly care for her infant in utero? First, she admits it to herself. Then, she seeks help. She may be surprised to find women in her peer group who have gone through precisely the same things as she has, or she may find help in mothering groups, or choose to engage the support of a counselor or psychotherapist. As will be discussed in "Gender Cultures" in Chapter Five, her best source of help for this type of issue is going to come from other women. They will understand her in a way that a man, even an understanding husband, simply cannot.

Touching the Unborn Child

The mother begins to touch the infant in utero. On one level, she is already touching the infant entirely from within her body. But since she cannot touch the infants' body with her hands or hold the child in her arms, her touch must become creative. Paula Matthew, practicing midwife for the past twenty-two years, explained, "Pregnant women who take the time to get alone with their babies and their bodies, and who enjoy the sensation of being pregnant, will usually develop a normal way of massaging their babies. The baby will push out, and the mother might push back, or playfully poke a bit, or rub a little foot, and in this way they develop a relationship." The infant feels the motions of her mother's hand, and begins to become familiar his mother's caress. Matthew said that when she doesn't see this happening—when she assists mothers who are uptight and excessively nervous about communicating with the baby in their womb, and who seem to feel very separate from it—that she becomes concerned about the future ability of the mother to parent her child well. Furthermore, through mother's touch on the outside of the womb, the infant is also touched by the mother's presence, interest and concern.

In addition, singing to the pre-born child, playing pleasant music for her, softly dancing with her, and taking long, relaxing baths "together" are only a few of the ways that mothers frequently get in touch with their infants before birth. Every mother-child relationship is unique, and thus it is up to each pair to explore together their new relationship in their own way. The mother who fears to risk this exploration may still give birth to a healthy, happy child, but she will miss out on the intimacy and joy of the love affair with her unborn child that is right at her own fingertips.

As vital as the mother's direct physical contact is with her unborn child, there remains an important aspect of "touch" that is less frequently acknowledged and understood among parents. This concerns the way the infant is "touched" by the environment that he is constantly exposed to, as well as the various impressions that he is at the effect of. To highlight this point, Dr. Sears points out a study revealing that some infants in utero show significant disturbance when their mothers attend loud rock concerts.[28] The same is true of violent movies. Similarly, if the mother has a habit of hanging out at bars where people drink, smoke and act in an aggressive manner, even if she herself is not drinking or smoking, the infant takes in not only the secondary smoke, but also the attitudes of the drunk and careless individuals in the bar. Given the fact that the individual adult's emotional network is largely at the effect of the impressions that he or she takes in from his or her surroundings, one can only surmise that the delicate and newly developing systems of the fetus are significantly more sensitive to this input.

The impressions that have the strongest impact come from the home environment itself—primarily, the mother's relationship with the father, or father-figure (if there is one). I often grieve for the children whose mothers allow themselves to remain in abusive environments, often justifying this by saying that they themselves are the target of abuse and not the children—particularly not the child in utero. This is like saying that the victims of a severe car accident are not hurt if they weren't driving the car. Where is the unborn child when the mother is getting abused by (or abusing) the father? Right there, feeling the constriction of the mother's body, receiving food from a body that is being biochemically affected by the abuse. And, it is not only a physically abusive environment that affects the child in this way. If the relationship between parents has a spoken or unspoken undercurrent of rage or resentment, the unborn child is at the effect of the invisible psychic "bullets" being shot between the parents.

This is not to say that a mother should stay hidden away at home for the duration of her pregnancy, or that she cannot have strong disagreements with her husband or go through a variety of positive and negative feelings about her pregnancy. However, since the baby is already influenced by her mother's circumstances, it serves them both if the mother is aware of this.

Touch is more than skin-to-skin contact. Touch includes a whole range of interactions between parent and child. Attending to the full spectrum of the needs of the child in utero are all ways of touching a baby.

Father of the Womb

The importance of the father's touch of his child in utero should not be underestimated. In his book entitled *Keys to Becoming a Father,* Dr. Sears discusses the various ways that the father can become part of the pregnancy, and thereby get "in touch" with his baby at this stage. Aside from the obvious need to help the mother in any way he can, both practically (e.g., assisting with chores, cooking, etc.), as well as through providing emotional support and understanding, the father can bond with the baby too. Dr. Sears suggests that the father talk with the baby. Reporting on fetal awareness research findings, he suggests that the baby can hear sounds outside the womb by the sixth month of pregnancy, if not earlier, and that the child actually may be able to hear their father's low-pitched voice more easily than the mother's higher-pitched voice.[29] (I would further suggest that the infant can feel the attention and presence of the father when he speaks to it even long before its auditory capacities have developed). Sears adds that when a father practices talking to the infant before birth, it is easier for him to talk with his son or daughter after birth.

Dr. Sears also suggests the "laying on of hands" for the unborn child. In other words, a father should touch the mother's pregnant belly, thus forming a relationship with his child. In doing so, a man expresses his affection and commitment to his partner as well as to their child. Paula Matthew has observed that infants in utero respond to the father's touch in different ways than they do to the mother's. She says that babies tend to calm more to a father's touch, whereas they respond more actively to a mother's touch.

Although these suggestions might seem silly at first, if the father, along with the mother, approaches this experiment with the attitude of "believing is seeing," he will soon discover for himself the delights of this new-found intimacy. When he touches his pre-born child, he imparts a small piece of life—his life—and that life is received.

Bonding is the Bridge

We have already established that bonding begins in the womb. The mother's very biology—her body and its cells and membranes and fluids—are inextricably connected to those of her infant. According to Pearce, however, this bonding must "then be confirmed and re-established after delivery." Therefore, the bridge between the safe world of the womb and the harsh exterior reality is the mother. It is her capacity to provide for her child the same qualities of security, nurturance, soothing and presence that he or she experienced in the womb, that will allow the newborn to feel a sense of belonging and safety in the world.

Ostensibly, it is her *relationship* with her child—which becomes well-established in the womb long before the child is born—that will serve as a matrix, or foundation, for the child. From this foundation he will eventually be able to venture out into the world, either securely or insecurely depending upon how well his mother has been able to provide for him. The bond that was present in the context of the womb will serve as a link, a center, a secure base from which to begin to perceive these new surroundings. When present, this link can root the child into a feeling of safety, security and confidence that will stay with him or her throughout the rest of life. Impeccable in its design, this process assists both the mother and child by initiating itself automatically in the ambiguous world of the womb, and then demanding that they recreate it in a concrete and seemingly separate world. The child's initial encounter with the world occurs in the chamber of his or her birth, and it is worthwhile to consider exactly what environment parents choose to birth their child in, and to consider taking some control in creating suitable circumstances.

The Birth Chamber

> *It is through our hands that we speak to the child, that we communicate.*
> *Touching is the primary language...*
> *The newborn baby's skin has an intelligence, a sensitivity that we can only begin to imagine.*
> *How, then, ought we to touch—to handle—a newborn baby?*
> *Very simply: by remembering what this infant has just left behind. By never forgetting that everything new and unknown might terrify and that everything recognizable and familiar is reassurance.*
> *To calm the infant in this strange, incomprehensible world into which it just emerged, it is necessary—and enough—that the hands holding him should speak in the language of the womb.*[30]
>
> - Frederick Leboyer

Many of today's children are born in large, busy and often impersonal hospital settings. Medicines, monitors and an array of doctors, interns and nurses assist the mother throughout the pregnancy and birth process, even though she inherently and intuitively knows what to do and how to do it. The mother's awareness of the importance that these first moments have for her infant, and her ability to trust this knowing and to place primary importance on the need to initiate the bonding process between herself and her child, is one of the most essential elements in the birth process. When this awareness is present—whether the child is delivered by a highly trained midwife, or born in an overcrowded hospital on the Lower East Side of New York City, is secondary. Primary to the child is his mother—her touch, her breast, her reassurance and her genuine welcoming of him into her life.

> *The core values and beliefs of both individual women and the society in which they live are condensed into visible, focused form in childbirth, where their perpetuation is either assured or denied. In the end—or the beginning—the salvation of the society that seeks to deny women their power as birth-givers will arise from the women who give that society birth.*[31] - Robbie Davis-Floyd

Whether the child is born in a hospital, a home or a swimming pool is not nearly as important to the child as is the attention, intention, and mood of the caretaker. The best the mother can give her baby is her body, her instinctual knowledge, her focus, her attention and her ability to touch. It is that which the infant will feel. In the following account, Sandra, a thirty-one-year-old mother, shares the story of her struggles to create a conscious and healthy birth environment for her child:

> I had been talking to my baby since conception. Yes, at first I kind of had to fake it—not really believing that she could hear me or feel my words, but I did it anyway, eventually understanding clearly that she could hear me. There was also an instinctual movement to always

be rubbing my belly. I think most mothers feel that, whether or not they allow themselves and their baby that kind of touch. It's not just the tactile sensation of skin-to-skin that the baby is feeling, or the sound of her voice that the baby hears, but the intention and the care of the mother. There was a longing and excitement for all of my senses to be with my unborn child—to touch her, to hear her cry, to smell her...

Finding out I was going to have a Cesarean was a blow to my pride—I wanted midwives, friends, a home-birth. I had an idea about what a conscious birth would look like on the outside and this wasn't it, but I decided to make my Cesarean work. I decided that I was going to make this as conscious a process of birth as possible, against all odds. The doctor, who was exceptionally open-minded, given his position, allowed me to have both my best friend and my midwife in the room. My friend's presence was crucial in the process—she held my hand and whispered to me what was happening the whole time. She midwived me while the doctors gave birth to my child.

When Kristy was born, the doctors stopped everything and immediately placed her on me. There in the operating room we had a minute of silence while my best friend said a prayer of welcoming to my child. Although this was unusual, the people in the room were visibly touched. The doctor later commented on how special and extraordinary the birth had been.

My friend accompanied my baby to the nursery and stayed with her while I was in recovery. There was an opening in the incubator for a hand, and my friend touched my child the entire time—I mean every minute. She didn't just have her hand on her while thinking about my recovery or her lunch—she was

> with her, and she's not the touchy-feely type either. I know that this affected my bonding with Kristy immeasurably, and knowing that she was being held helped me a great deal with my concerns about her not bonding properly. My friend explained to my baby what was going on for me and that I would be with her soon. By the time they got to me, she and my baby were old pals and have been ever since. A lot of mothers think that if they can't be there that bonding can't happen, and it's rare that someone else is willing to be there in that way, though sometimes fathers are. Nonetheless, although it is preferable for bonding to be with the mother, what is most important is that the baby bond—period.

From a physiological perspective, Ashley Montagu says that the first thirty minutes are the most crucial in terms of bonding, as bonding at this time enhances the essential hormonal changes that will contribute to the child's optimal functioning.

Sandra continued:

> *When they brought her to me a few hours latter, I picked her up, ripped off the blanket and she immediately began to breast-feed—that was the beginning of nursing. I know many of the doctors thought we were foolish, but it didn't matter. My child is now three years old, is bonded, and is happy—that's what matters.*

The midwife who attended the birth similarly suggested:

> *I think that at first the doctors are a bit taken aback when somebody like Sandra comes in and wants to*

know everything, wants to be a part of everything, and wants to hold her baby immediately after the operation. But after the initial surprise, they like it. It makes them feel more human—it makes us all feel more human. It's nice for them to have the opportunity of being plummeted into that kind of intimacy so that they can experience it without having to prescribe it for everyone. They get a chance to be there and be a part of a healthy birthing process. They feel the difference.

Sandra's story illustrates that if the child is really put first, one can be victorious against all odds.

Whatever environment parents choose, the consideration of bringing a midwife or a woman friend who is experienced with children into the birth chamber is a worthwhile one. (The father, of course, is a welcome and valuable presence, but what he provides is distinct from what the woman helper is able to provide.) A woman's intuition is never closer to the surface than at the birth of a child. Oftentimes, childbirth is induced by "force"—involving drugs and intrusive medical instruments that could otherwise be avoided. When possible, a natural birth is more gentle on the infant's delicate system, and is empowering to the mother as it allows her to be more fully present for the birth of her baby. An advantage of homebirths, but also possible during hospital births as well, as evidenced in Sandra's story, is that there tends to be a more open attitude toward allowing women's knowing to play a key role in the birthing process. Rachel, a mother of two who has attended numerous births, shared this story:

A close friend invited me to be one of her birth attendants. I was delighted, like always, because there is a

> *specific, raw, primal energy present at a birth that is almost addictive. Anyway, her labor was going on pretty smoothly, but the baby was not crowning. I asked the doctor what was going on and he said, "I think we'll need to use forceps." I communicated to the mom that this is what the doctor was planning. She asked me if there was anything she could do to avoid this. I said to her, "We need to get the baby out." I told her how to use her breath and a specific visualization, and within three breaths her daughter was crowning and came out shortly thereafter. The raw power of woman was evident to me during that birth—that by her will, whatever it was that was holding the child back, she was able to overcome and have the result be successful without medical intervention.*

Many babies do not come out readily and it is common for babies to initially have difficulty with breathing. For this reason, doctors have a variety of methods at hand to encourage the first breath—ranging from turning babies upside down and hitting them, to suction machines to remove the mucous from their lungs. However, one of the most unrecognized and understated tools in the doctor's reach is physical touch. One female medical doctor who is aware of the value of touch suggested, "I've seen babies born who are blue or gray and as soon as they are touched there is a pinkness that emerges at the heart and spreads throughout the body." Another obstetrician-gynecologist who practices alternative women's healthcare added, "When we have babies that are slow to breathe, if we just rub them and touch them, almost always that's all that it takes." The following story exemplifies this point more immediately:

I was a backup nurse midwife and called in late on the scene because it looked like the doctor wouldn't be there. When the baby was born, we went ahead and put her up on the mother, but the main doctor arrived and wanted to see if the child was doing okay so she took the baby from off the mother's abdomen and right away the baby stopped breathing. An assistant listened to the baby's heartbeat and reported that it had stopped, so the doctor started doing very aggressive CPR on the baby. I asked if she could stop for a minute, or if she could do the CPR while the baby was on the mother's body. She stopped, we put the baby on the mother, and the infant resumed breathing as normal. It seems so obvious, but I guess it's not.

In *Nurturing the Unborn Child,* Verny tells a similar story about a Seattle pediatrician who was treating a very ill baby in the intensive care unit of the hospital.

Hooked up to a battery of life-support machines, the baby was flooded with light and exposed to more sound than you might hear on Fifth Avenue at rush hour. The child was just not getting enough oxygen, and he was turning blue. The doctor decided the infant was going to die anyway, so he took him off life-support systems, shut off all the machines, and turned off the lights. Then he took the baby out of the crib and rocked him in his arms. Within a few minutes the baby turned pink, and his recovery was complete.[32]

Hospital Birth v. Homebirth

For those who are aware of the acute sensitivity of the newborn infant, its need for touch, and the tendency for hospital births to be highly impersonal, the consideration of hospital birth versus homebirth, and medical doctor versus midwife, is a valid one. To put the dilemma simply, doctors concerned about being sued for malpractice are very uncomfortable having their patients give birth outside of a hospital setting in which the baby cannot get immediate medical attention if any difficulties arise. They are also angry about the last minute complications that have come up in natural births—like when the baby is rushed to the hospital because of difficulties during the homebirth, and the doctor at the hospital is held accountable for the procedural difficulties, even if it is the first time he has met the mother and baby. Furthermore, most doctors are not well educated in the science of midwifery. Popular stereotypes dictate that midwifery is outdated, idealistic and irresponsible. Women who choose to have their babies birthed with the assistance of midwives are often not taken seriously by medical doctors.

Parents also struggle with what Vancouver gynecologist Dr. Karen Buhler calls, "the where question." The foremost concern of most parents in deciding whether to have a hospital birth or a homebirth is the safety of their child. In a hospital, the safety of the birth is taken for granted. Technology, lifesaving treatments and highly trained personnel help to assure the parents of a safe birth. On the other hand, parents know that a hospital birth means the possibility of unnecessary intervention, and also a less personal environment where the parents have less control about the birthing process.

On the flip side...there is no flip side. Yes, there are midwives who have been irresponsible. Yes, there have been unforeseen situations that have arisen in homebirths that required immediate medical attention. Yes, there are many instances, like Sandra's, in which a hospital birth is required. Still, the arguments against the practice of midwifery and homebirthing are generally invalid. In fact, some studies on the subject reveal that infants tend to have as good a chance, if not a better chance, of survival when birthed naturally than when born in a hospital. In her book, *Homebirth*, educator Sheila Kitzinger pointed out that in 1986, the peri-natal mortality rate in the Netherlands was 13.9 per thousand in hospital births, compared to a mortality rate of 2.2 per thousand in homebirths.[33] Buhler further elaborated:

> *Contrary to popular belief, homebirth is safe under the right conditions. In fact, in a well-selected, low-risk population assisted by competent, experienced attendants with good access to hospital for transfer if necessary, home birth is as safe or safer than hospital birth.*[34]

In 1993, the British House of Commons Select Committee on Maternity Services, in a report called "Changing Childbirth," emphasized:

> *Women should be given unbiased information and an opportunity for choice in the type of maternity care they receive, including the option—previously denied to them—of having their babies at home or in small maternity units...the policy of encouraging women to give birth in hospitals cannot be justified on grounds of safety.*[35]

Mothers who choose the option of homebirth or birth with a midwife have usually given the topic of birth ample consideration. All mothers want to give birth to healthy babies, and because of the prevalence of the societal belief that hospital births are safer, no mother chooses the option of homebirth without first facing this obstacle. I have spoken to many women who have said that when they had their first child they were in their late teens or early twenties and really did not know they had options. Mothers who choose homebirth are often given more prenatal care, support and education, as the midwife is often a partner in the birthing process, which includes pregnancy and all the biological and emotional stages that the mother goes through in preparing to give birth. Furthermore, midwives tend to be sensitive to both common and uncommon cycles of labor and delivery. Because midwives have a greater freedom to trust the woman's natural cycles of labor than do most medical doctors who live with the ghost of a malpractice suit on their left shoulder, they are less prone to intervene with unnecessary medications and anesthetics. While medical doctors clearly provide a certain kind of security during childbirth, they do tend to interfere with the very intimate process that is occurring between mother and child.

Natural births are safe. "Since the 1920s," reported Janet Issacs Ashford in *The History of Midwifery in the United States*, "nurse-midwifery services have achieved lower-than-average infant and maternal death rates...this holds true despite the fact that nurse-midwives have primarily served low-income women who, demographically, are at higher risk for complications." [36]

Can a young woman, seemingly psychologically unprepared for the immense responsibility of mothering, effectively take charge of her own labor? Paula Matthew, from her experience delivering over 950 infants, says yes.

> *When I was just beginning my career as a midwife, a neighbor called me in to assist with her daughter's birth as the mother had gotten very nervous. The girl was fifteen. It was a real eye-opener for me to see the control she had. She made all the decisions about what she would do and when; she called the shots about who would get to hold the baby, who would get to touch it, what it was going to wear, who would get to wash it—the kinds of things that to a medical practitioner would probably sound very meaningless, but to a mother and child can make all the difference in the world. This fifteen- year-old was there saying, "This is how we're going to do this, this is how we'll do that...Give me my baby now." Everybody was there to assist her with what she needed—she may have been fifteen, but she knew exactly what she wanted and exactly how to mother this child.*

Matthew contrasted this to her own experience:

> *When I had my first baby at age eighteen, a nurse actually called me an abusive mother because I sat up wanting to hold my baby right after she was born and wanting to breast-feed her. They were telling me that I was too young to know how to take care of my child, that I would be an abusive mother because I wasn't together enough to know what to do.*

Similarly, Judith, a mother and former educator in the Summerhill educational programs, reported:

When my first baby was born we stayed in the hospital for four days, seeing one another for only a couple of hours each day. That can sure break the bonding process! I was twenty-one—they drugged me for the birth and told me I was an ignorant mother. I hadn't studied anything and my own mother had done the same. I wasn't awake when my baby was born...they took him away to the nursery. We still miss something and we both know it.

A co-worker reported to me a startling example of unconscious birth that happened to a friend of hers:

She was given an epidural (spinal anesthesia) for her delivery, which is a common practice today. The woman had a history of rapid deliveries, so the nurses had instructions to check on her every fifteen minutes. At one point they went in to check on her and noticed she was reading a magazine. However, when they looked again they saw that the baby was already out to its knees with the cord tight around its neck. She had delivered and didn't even know it—she was reading her magazine.

No woman who has looked within herself to even the most cursory degree could tolerate that kind of unconsciousness in her child's birth chamber.

From the perspective of touch, hospital births may be very rough on the infant. His first experience of touch is usually either the doctor's gloves pulling on his head, or metal forceps grabbing him. (There are times when this needs to be done, but as in the previous example of Sandra, the mother can be communicating to the infant about what is happening, assuring him that whatever procedure is going on is for his benefit.) After a momentary encounter with his mother, he may be whisked into a cold room, and placed in a warmer where nobody is going to touch him. Eventually, the baby is returned to his mother, who, in this day and age, will often not even offer the nurturing breast that the infant knows, instinctually, he must find.

Bear in mind, from what we have discussed in the beginning of this chapter, that the infant who is going through this process is totally conscious. He does not have a language for this experience, but is nonetheless fully aware of it, and will experience the long-term effects of it throughout his lifetime.

For a peek at the other side, consider this scenario. The parents have planned for this child well in advance. They have already formed a relationship with the infant in utero, and want his first moments and hours of life to be full of gentleness, and as welcoming as possible. They have chosen to have the birth at home, and when the mother begins labor, the doctor or midwife whom they have engaged for this purpose is called, and also a special friend whom the mother has chosen for support. The room has been prepared, and is warm, with dimmed lights. Plenty of blankets are available so both the mother and the newborn baby will be warm and comfortable. Perhaps the parents have chosen to have music playing softly in the background. During her contractions, the mother is encouraged to relax and follow her breath. Her pain is respected. She wants to be fully present without anesthesia so that she can be there with her child, in mind and in body, feeling the slightest movements of birth. She has been taught well in advance that there are many positions in which to give birth, and she changes positions as she desires.

The doctor or midwife delivers the baby and place the newborn immediately on the mother's abdomen covering him with a blanket. All in attendance are sensitive in allowing the child to be silent or to cry, to nurse or to simply lay there. They naturally touch the infant, massaging him gently. Eventually, the umbilical cord is cut, the vernix is rubbed into his skin, and then the infant is returned to his mother for rest.

One woman described these moments vividly:

No matter how exhausted a mother is after the process of labor, when a child emerges from the womb and she can see him, touch him, or reach between her legs with her own hands and pull him to her...the magic, the light, the revitalization is amazing. The mother may be asleep, eyes half-closed, but as soon as the child is seen or felt, it's like dawn. It's an exhilarating brightness. The mother's face lights up. It is the most unbelievable, awe-producing event. There's such a surge of energy, aliveness, awakening.

Ultimately, the issue of home versus hospital birth is not a "good-bad" situation. It is a personal choice. Karen Buhler, M.D., suggests:

My advice as a doctor to parents considering the "where" question is that you first take some time to formulate what kind of birth experience you hope to have, what your philosophy of health is, and what services you prefer. Then choose a caregiver (or caregiving team) that you trust and feel good with who shares your beliefs and goals. The caregiver can assist you by providing information, assessing your risk factors, and describing the options and facilities in your area. She can visit your home and help you assess its suitability for a home birth.[37]

Again, the issue is touch—touch being the mother's loving regard for the infant even while he is still in utero; touch being the intention of the mother and her attention on the child; and touch being the unselfish physical affection she adorns her child with throughout his childhood.

THE THREE B'S—BREAST-FEEDING, BACKIN' DA BABY AND BEDTIME

The life-giving body of the mother is already coded with everything it needs to know to care for her offspring. The enactment of the three principles described below catalyzes this organic knowledge in the mother, allowing her to provide fully for her child.

The First B—Breast-feeding

When a mother makes the choice to breast-feed her child she is acknowledging the perfection of nature—that she has been given the perfect substance with which to nourish her child: a substance far superior to any processed formula.[38] Whether or not she is able to justify it verbally, she knows that this is what her child wants, and *needs* from her. Thus, she takes a stand. She recognizes the societal bias against breast-feeding; a bias which is largely a function of the multi-billion dollar corporations which promote the production of baby formulas and bottles. She is aware of how contemporary culture often scorns the breast-feeding mother, and wants her out of public view. She decries all the advertisements showing healthy babies who drink only bottled milk. She knows her own body—how the milk flows naturally to her breasts—and how her child desires to suckle and mold himself into her. She knows her own desire to hold her child closely to her skin. A letter from *Mothering Magazine* reads:

> *Breast-feeding is more than pumping nourishment; it is a psychic bond, which is not disposable from nine to five.*[39]

It is apparent that breast-feeding offers the best possible beginning to a child's life. Some mothers, however, are unable to provide this for their child, due to a wide range of physical, circumstantial or psychological reasons. A dedicated, devoted mother who places primary importance on touch, will bond with her infant regardless of whether she bottle-feeds or breast-feeds. Montagu says:

> *In the bottle-feeding situation, which is the rule in America, the infant experiences the very minimum of reciprocal tactile stimulation. The deprivation of tactile stimulation experienced in this way by both infant and mother explains the institutionalization in American culture of the nonexpression of affection, especially between mother and baby, through close physical contact. Tactile contact between the American mother and child expresses caretaking and nurturance, rather than love and affection...*[40]

Breast-feeding not only provides physical nourishment, but the opportunity for mother and child to bond. Leaving her womb, the closest the baby will ever come to his mother again is while nursing. When the mother is present with her nursing infant, the infant will often hold unwavering eye contact with her. The mother is there for her child, providing nourishment, fulfilling his hunger on so many levels. When she touches her infant in this way, both through her eyes and her body, she is communicating to him or her, "You are safe. You are wanted. You are fed. You are loved." She doesn't need to give him excessive toys or treats. *She* is what the baby most wants and needs.

The conscious mother nurses her child according to the child's desire to be fed. Understanding that the child seeks her breast for many sources of "food"—bodily nourishment being only one of them—she does not impose her ideas and preferences about when her baby should be hungry and need to eat. The child knows when they need "food," and communicates this to the mother by nuzzling into her breast, reaching to her, or crying out in a particular way. When the mother responds to this consistently, she empowers her child to trust his or her own sense of what they need. Her actions demonstrate that *the child's basic needs will be met.* (Many adults overeat, smoke addictively, and so forth, still trying to prove to themselves that their basic urges will be satisfied.)

Breast-feeding mothers are often startled to find that their child wishes to nurse far longer than they had imagined to be "normal." Although practically unheard of, and nearly taboo in American culture, it is common in many, many cultures throughout the world for the infant to continue to breast-feed for two, three, four or more years, or until a younger child is born. Although the modern mother may be fearful of breaking a societal taboo, given what the costs are for an unbonded infant, breast-feeding over an extended period of time is worth considering. Regardless, whether she nurses her child for one month or two years, however long a mother is willing to breast-feed her child will be felt throughout the child's life.

Beyond this, breast-feeding is a crucial factor in the development of intelligence. Pearce suggests that chemicals found in the mother's milk are actually necessary to stimulate mid-brain functions that regulate certain aspects of intelligence.[41] Some studies have revealed that breast-fed babies grow up to be more intelligent than those who are bottle-fed. In 1992, Alan Lucas from the Dunn Nutrition Unit in Cambridge, Massachusetts, published a study showing that pre-term infants who were tube-fed breast milk scored much higher on developmental tests than babies who were tube-fed formula."[42] Similarly, researchers at the National Institute of Health conducted a study in which they followed 855 children from birth to school age to measure the correlation between breast-feeding and intelligence. They found that the longer a child had been breast-fed, the smarter and more coordinated he or she was.[43]

The Second "B"—Backin' da Baby

> *Every nerve ending under his newly exposed skin craves the expected embrace, all his being, the character of all he is, leads to his being held in arms. For millions of years newborn babies have been held close to their mothers from the moment of birth. Some babies of the last few hundred generations may have been deprived of this all-important experience but that has not lessened each new baby's expectation that he will be held in his rightful place.*[44]
>
> \- Jean Liedloff

When I initially began to befriend the Creole-speaking Jamaican immigrants who settled on the

Atlantic Coast of Costa Rica (where I was working at the time), understanding their language, although a derivative of English, was nearly impossible. Often I would inquire as to the whereabouts of a child from a sibling or the child's father. "Oh da' mudda she backin' da baby," I would often be told. Timid to ask for a translation, I began to notice over time that infants were constantly held in a sling on the mother's back, or perhaps in her arms, for the first several months of their lives. At other times they would be sitting on her lap, crawling near to her, or in the arms or lap of the father, sibling or grandparent.

In all of my extended visits, as well as during the time I spent with the indigenous people of the forest who lived nearby, I never saw anything that resembled a crib, playpen, walker or stroller, much less a leash! (To imagine a BriBri Indian woman navigating the paths of the rain forest with her machete in one hand and pushing a baby stroller with the other would be absurd!) Children were raised "in arms," to use a term coined by Jean Leidloff in her revolutionary book (in terms of contemporary Western culture), *The Continuum Concept.* "In arms" means that the infant is constantly in contact with her mother's body or that of a primary caretaker's body for the first several months of life, if not longer.

This constant contact is the second essential aspect of the bonding process. Not only is the tactile stimulation from the constant movement of the mother's body enlivening to the growing infant, but the young child becomes bonded into the family, and becomes integrated into the world around her, simply by being there with the mother as she goes about her daily chores and interactions. The mother is aware of her child and attends to her needs, but is not excessively doting over her. Instead, she brings her into the full spectrum of life experience, allowing her to simply be a part of it.

This extensive contact in early life gives the child a sense of security, a certain feeling of "home" that will remain with her throughout her life. The child will automatically develop a sense of confidence in her own body. Leidloff explained, "If he (the baby) feels safe, wanted, and 'at home' in the midst of activity...his view of later experiences will be very distinct in character from those of a child who feels unwelcome, under-stimulated by the experiences he has missed, and accustomed to living in a state of want, though the later experiences of both may be identical."[45]

I was recently telling a four-year-old child about life in that particular village in the rain forest where I worked, and how each of the houses was a one to two hour walk apart from the next one. "But they have roads to drive around in?" he asked.

"No," I told him, "The growth in the forest is too thick and the people cannot afford cars."

"Then they must have good sidewalks so the moms can push the kids around in strollers," he pressed.

"No," I continued. "Until a child can walk, he or she is carried, and once able to walk well (occurring usually by age two to three), the child walks along with everybody else."

This child was baffled. He probably could not imagine such autonomy, capability and confidence on the part of a young child. What I did not mention is that not only do these young children walk, but I would often follow them from far behind, laughing at my own clumsiness as they ran, seeming to fly, down the mountain, avoiding roots, lizards, branches and even an occasional snake, with the precision of a hawk.

These children, held in arms and nursed until they were ready to move outward, were totally competent. At first I was surprised to find four- and five-year-old children from across the forest showing up at my doorstep without their parents, but I quickly understood that they were already knowledgeable of all the dangers of the forest, that they had memorized all of its windy paths, that they knew how to protect themselves and where to get food and water. These were very capable children who were trusted by their parents.

Author Barbara Aria and researcher Carroll Dunham came to a similar conclusion:

> *Studies in Uganda have shown that babies who are carried in upright positions are quicker to walk, and develop faster in other areas too; the upright position heightens a baby's visual alertness while developing muscles in the back and neck. Carried around all day, babies become familiar with their worlds as they watch from their secure vantage point. Because they're held close and upright they stay calmer—studies show that they even cry less than babies that aren't carried regularly.*[46]

Backin' da baby does not suggest that the infant must be carried only on the back. Front, side, right, left...in hoods like Eskimos...the variety of positions and types of attachments are numerous. What the baby craves is contact. Mothers can learn to vacuum, cook, clean, shop and work with the baby in their arms, though it might take longer. Many mothers in the West are enjoying a rediscovery of the simple sling, which they can take on or off with ease, and in which a baby may sit or lie as they please. Furthermore, as the child matures month by month, she will naturally want to venture out from her mother's body. "In arms" naturally becomes "in view" as the child gains confidence and curiosity about her surroundings.

When deciding how long to keep the child "in arms," the key is the child's timing. The child is not placed in a walker prematurely because there is no need to encourage him or her to walk before their body is ready to. This was certainly true in my own case, as I was born with a displastic hip and was carried around for the first two years of my life with my legs encased in a large metal brace. The day the doctors took it off, I was able to walk as any average two-year-old could.

When a child is bonded and feels secure, her physical skills and intelligence develop naturally. No external timeline for growth is necessary. Each child bonds in his own time, so while one can generally estimate how long the child might wish to nurse, or in what way or for how much time the child will want to be carried, the bonded mother-baby team will find each stage self-evident.

Prescott summarized years of statistics and scientific research on this subject in the following:

> *I have concluded that the single most important child rearing practice to be adopted for the development of emotionally and socially healthy infants and children is to carry the newborn/infant on the body of the mother/caretaker all day long, where the continuous "backpacking"...becomes the best "behavioral vaccine" against depression, social alienation, violence and drug/alcohol abuse and addiction in later life.*[47]

The Third "B"—Bedtime

> *...The room that filled with suspicion at night: you made it harmless; and out of the refuge of your heart you mixed a more human space in with his night-space. And you set down the lamp, not in that*

darkness, but in your own nearer presence, and it glowed at him like a friend.[48] - Rainer Maria Rilke

"When I told my family that I didn't want them to buy my two-year-old a crib because she was sleeping in the bed with my husband and I, they were scandalized," began one mother. "My brother and sister were shocked too—'You can't be doing this!' they said."

All over the world parents and their infants sleep in close proximity with one another. In the Yucatan Peninsula of Mexico, where people commonly sleep in hammocks, one can purchase a "family-sized" hammock which will amply accommodate the parents as well as one or more children. In Tamil Nadu, India, the whole family sleeps on large palm mats on the floor, or on several cots piled into one small room. Japanese infants share their parents' futons.

When I was a young child there was nothing as reassuring as my mother's bed. Terrified of the night, I vividly remember lying in bed hour after hour, watching the fluorescent blue lights of the digital clock by my bed, thinking of my mother sleeping soundly in the next room. When the sleeplessness went on too long, the fear too much, I would quietly get out of bed, tip-toe down the hallway, and slip in my mother's bedroom through the door that was always left open a crack. As I crawled into bed next to her, and she spooned me into her body, I would fall asleep instantly.

Coined "shared sleeping" by Dr. William Sears, and "the family bed" by Tine Thevenin, the concept of parents and their infants sleeping together is far from novel. Yet in the West, this concept is just beginning to be renewed. For the past several generations, if the family can afford it, the American baby is given his or her own room from birth, complete with crib, hanging mobiles and toys, and perhaps an intercom monitor connected to a speaker in the parents' bedroom if it is too far away for the mother to hear her baby's cry. Although parents might do a bedtime ritual with their young children, and children who are unable to sleep may occasionally crawl into their parents' bed late at night, to most people the idea of a "family bed" is preposterous.

It is useful to consider why this idea disturbs us so much. I suspect that, like so many unfamiliar ideas, it is upsetting simply because it is unfamiliar and new, and not because of anything to do with the shared sleeping arrangement. When living in India, I was invited by a very wealthy family to the marriage of their son. Late at night we returned to their home, which had complete plumbing, high ceilings, classic art on the walls—all very rare in India—and several straw mats on the hard marble floors where all the women and children in the extended family were to sleep. At first I was taken aback, for I saw that I was to sleep with sixteen women and children. How would I tolerate the noise? The lack of privacy? Furthermore, I did not know exactly what to do—should I change my clothes? Should I ask for a blanket? Whom should I sleep next to? Should I just lay down and close my eyes? Yet, soon enough I saw everybody find a place, lay down fully dressed in their *sarrees,* and go to sleep. I followed. Once I dropped my idea about how strange it was (especially for a wealthy family, I arrogantly thought), I found that I greatly enjoyed the intimacy of the experience.

The hours of sleep are very vulnerable times for all people, especially for children. Even while sleeping, people are aware of where they are and how safe or unsafe their surroundings are. For example, a couple who sleeps together is aware of one another's presence, and their sleep patterns often change in the partner's absence. Similarly, the child is aware when the parent is there, and for many children this awareness provides a much-desired sense of safety and security. The mother's touch provides a quality of soothing that nothing else can. When the child feels this security, he or she is freer to fall into deeper states of relaxation, rest and exploration of the dreamworld. Pamela, mother of two, suggests:

> *The family bed can be tried as an experiment. Most moms are instinctively in touch with what their infant needs, only we have become separate from it. Whereas once I could not have imagined the idea of the family bed, now I can't imagine life without it—especially for my children...At night after I tell my three-year-old a story, she gets on top of me and falls asleep on my chest. Then when she's sound asleep I lay her beside me. Having done this, now when there's a strong emotional upset or distancing between us, there is something about laying on top of each other with our hearts touching that serves as a way of coming back to center and grounding.*

Peace of Mind

Many parents worry about a disorder in infants known as SIDS, or Sudden Infant Death Syndrome. Although 99.8 percent of all babies don't die of SIDS—the stories that parents have heard about infants under six months of age dying suddenly in their sleep, apparently having stopped breathing for an undiagnosable reason, has brought considerable concern to many. Pediatrician Dr. William Sears was initially confronted with this disease when the mother of one of his three-month-old patients called him crying that her baby was dead. When Sears found that he had no answers with which to console her, he took it upon himself to research a cure for this disease. After years of study, his firm hypothesis is as follows:

> *I believe that in most cases, SIDS is a sleep disorder, primarily a disorder of arousal and breathing control during sleep. All the elements of natural mothering, especially breast-feeding and sharing sleep, benefit the infant's breathing control and increase the mutual awareness between mother and infant so that their arousability is increased and the risk of SIDS decreased.*[49]

Similarly, research conducted by James McKenna, an anthropologist at Pomona College in California, suggested that shared sleeping may reduce the chance of SIDS.[50] Once again, I believe that science will prove to us what our logic already knows: if the mother is deeply bonded with her infant, and is there for him body-to-body as he sleeps, the infant will breathe more easily and she will be more aware of his needs during the night should problems arise.

In planning their sleeping arrangements together, it is helpful for parents to pay attention to their children's needs. Some children want to sleep between both parents, whereas others prefer to sleep "spooned" into the mother. As children get older, they may wish to spend part of the night in their own bed or their own bedroom, and then come to sleep in their parents' arms for the remainder of the night.

Parents are often concerned that shared sleep will mean they won't get enough rest. Although there is an initial period of adjustment, preliminary studies have confirmed sleep-sharing mothers did not get less total deep sleep.[51] Furthermore, the peace of mind that parents long for more deeply than a peaceful night's sleep will arise when they become aligned to the genuine needs of their children, and are able to provide this for them.

Sears concluded, "Based upon my own research, 25 years of pediatric experience, and 17 years of sleeping with our babies, I have come to the conclusion that the safest place for babies to sleep is with their mothers."[52] As with the child who is breast-fed and the one who is carried, when the child feels sufficiently secure and ready to move on, he or she will request his or her own sleeping space.

INFANT MASSAGE

Another wonderful way that parents bond with their babies is through infant massage. Although not always called by name, this practice is engaged by mothers all over the world. It is natural for a mother to massage her baby, invigorating the child's skin and thereby stimulating the organs, and providing the child with comfort and pleasure.

Massage with infants serves many purposes—both physiological and psychological. In her article, *Massage Therapy for Infants and Children,* Tiffany Field suggests that infant massage has been documented to: 1) decrease the production of stress hormones, 2) reduce pain associated with teething and constipation, 3) reduce colic, and 4) help induce sleep.[53] It also serves many other functions including stimulating the baby's organs, helping the nervous and digestive system to function more optimally, increasing daily weight-gain in premature infants, and inducing more responsiveness and improved sociability. Some reports indicate that infants who are blind and/or deaf become more aware of their bodies and that infants with cerebral palsy also benefit by more organized motor activity.[54]

The greatest benefit of massage, however, is the unparalleled experience of soothing, nurturing, safety and pleasure that only loving touch can provide. The child's body is teeming with life and fully open to receiving its pleasures, untainted by the emotional blockages, environmental numbing and body shame that will inevitably accumulate throughout his or her lifetime, making receptivity to touch more difficult. The infant has not yet entered the world of concepts, ideas and protocol, and can best be communicated to through the means of touch.

Furthermore, massage is likely to nurture the parent as much or more than it does the infant. Some parents are afraid of their children—afraid of their young and vulnerable bodies. Other parents are afraid of their own sensual desires to touch their children because they fear that their touch is sexual. Some are afraid of allowing their babies the pleasure that they (the parents) may have suppressed in their own early experiences. Massaging their infant often helps both mothers and fathers to extend themselves to their children in ways they would otherwise be unable to, thus bringing them into *relationship* with their child through the vulnerable and deeply communicative means of touch. Massaging also allows parents to know their child's body better, to impart to the child that touch is good, and that there is safety in receptivity. Again, it is the touch that matters—feelings of inadequacy or fears due to a lack of training can be coaxed into the backseat as we place our primary attention on this natural, health-producing practice of massage.

Children *need* touch. Providing them with ample, nurturing touch in infancy sets the foundation for a future life that will be full of intimacy and self-confidence. Yet the child's need for touch does not end when infancy fades. Touch is important throughout the child's growing years—

important for both child *and* parent. Throughout childhood, the vehicle of touch allows children and parents alike to demonstrate a sense of appreciation for one another that nothing besides genuine touch can provide.

CHAPTER 3

Second Touch: Affection with Children

When my Scottish sister-in-law's father finished reading my first book, he put it down and said, "My mother never showed me any affection when I was a child so I spent the first ten years of my life thinking she didn't want me." Children don't stand a chance of sanity and peace of mind in adulthood if they are not given proper affection in infancy and childhood. Whereas the most critical period in terms of touch is the first moments, hours and years of the child's life, his or her affectional needs will continue throughout the formative years. The initial bonding is still vulnerable, not yet rooted and stabilized, and can therefore be undermined when touch is not consistent as the child matures.

For many years I wondered just what a bonded child would look and act like? I have attempted to observe the differences between those children whom I know have been raised with ample touch and affection, and those who have not (not because they have uncaring parents, but usually because the parents are rigid, controlling or simply too wounded themselves to be able to raise their child in a healthy, affectionate environment). What I have discovered—at least at first glance—is that the bonded child looks and acts like most other children. He or she eats, sleeps, plays, spends time with friends, cries, laughs...the whole human roller coaster of experiences and emotions. However, a closer study of the bonded individual reveals a quality of acceptance—both of himself as well as of others—that is very rare. His life is not based upon a search for approval and validation from some ambiguous source. His interactions with others are characterized by the absence of insecurity, fear and the wish to impress others that is so common. You sense that he does not put his personality on the bed table when he goes to sleep—that what you see is what you get, and that this is perfectly all right with him. Furthermore, the bonded individual tends to be kind and considerate of others. At his center he knows he is loved, and therefore has no fort to defend, and wears no suit of armor to protect himself. Kim, an educator, shared:

My daughter married a Mexican man who has a great extended family. They love the kids. The kids are wanted and no matter what they do the adults are affectionate. The men are comfortable walking around with their arms around one another and hugging. My grandson is also being raised in this way—it feels real healthy. At his birthday party there were sixty people and you would see him going around from one relative to the other, climbing on their laps—he knows them all by name. He picks the birthday cake up in his hands and it's O.K. that he picked his nose before touching the cake. Nobody cares. Nobody says a word. There's no message at all of not being accepted for the things that come naturally to children, so the kid grows up with a sense of being O.K. with himself. I look at my son-in-law and think of how very rare it is to see a young man in his mid-twenties who feels good about himself and has no problem saying so. He has self-confidence while at the same time is capable of extending warmth and of putting his wife's and his baby's needs first. He can feel good and important without feeling superior. It is very healthy and very balanced.

Basic sanity—this is the gift that awaits the bonded child.

HOW WE LEARN TO LOVE OUR CHILDREN

Children need affection, and lots of it. Everyone would agree that all children need love, but love is a concept that is overused and under-practiced in the modern world. We as a culture need to become more specific about what this "love" means and about how it shows up in our interactions with our children. It might serve us to temporarily forget about the notion of loving our children, suspend our belief that all the strong emotions we feel about our children are actually love, and begin to touch them, to hold them, to be there for them in ways deeper than words can express and basic caregiving can offer. "I am here for you and I respect and honor you," is what needs to be communicated to the child. To truly respect and honor our children and to give them the touch they need demands a great deal more from us than our commonly held notions of love do.

Essentially, children learn what it means to be affectionate from their mother and father. They also learn true masculinity and femininity from them. Children come to trust or distrust all men and all women, and to relate compassionately or defensively with them, by how their mother and father touched or failed to touch them. Although nannies, baby-sitters, older siblings and staff workers at daycare centers can be important role models for children concerning how and when to touch others, and what it really means to be a man or a woman, nobody can model this for them as effectively as their own mother and father can.

We teach children to demonstrate spontaneous affection by modeling that in our own lives. Ironically, children learn more from the way that adults express themselves with one another than they do from how the adults express affection to children. When parents express affection and love to their children, but treat one another as less than the dirt on the bottom of their shoes, the children know that something is not right. These children cannot fully trust the affection that is given to them because they see that somebody else—who Mommy or Daddy supposedly "loves"—does not receive the same regard. The child concludes both that the affection they are receiving is not real (i.e., "Mommy isn't really a loving person, she is just nice to me because I'm her kid"), and also that affection can be withdrawn at any point—that it is conditional. When these children become adults, unless they have undergone a thorough process of self-examination, they will demonstrate affection in their lives in the way they saw their parents model it with one another.

Whereas some adults may feel that they do not need or even want touch, all children crave touch, and they experience a lack of touch as traumatic. In the rare instance when a child appears to have an aversion to his or her parent's touch, trauma has already occurred, if not during the initial period of bonding, than in the womb or through some other event in the child's early life.

My client Janis recently commented that for the first two years of her life she would not allow her mother to touch her. Aware of the needs of all infants, I was surprised by this. She then went on to say that at the age of nine months she was left at home with a baby-sitter for three days while her mother went on a short vacation. When her mother returned home, Janis was not there, and a sheepish baby-sitter pointed her to a neighbor's house. Apparently, the baby-sitter had left Janis wailing for hours at a time, and the neighbor, hearing her cries, had come over and insisted upon taking the child to stay with her until the mother's return. Although her mother felt terrible about what had happened, the damage was already done in the hours of screaming and sobbing in which Janis had not been attended to—the hours of betrayal in which she did not know if her mother would ever return. I later questioned Janis about this, as she had so often spoken of the closeness between she and her mother. "Oh yes," she replied, "but I would only let her get *so* close. I put up a wall in a very specific place and there was no way she was ever going to get through that again."

Another function of the parent's touch is to help the child to process emotions and energy. When a child is upset, it is not necessary to "coo" over him or her. Simply by taking a shaking child into her arms, the mother imparts a steadiness that is reliable. The parent's non-smothering touch communicates to the child, "You're O.K. I am here for you." Such touch allows the natural cycle of pain to move through the child—neither encouraging it's indulgence by excessive pampering, nor discouraging it's expression by denying the child his right to cry. The child then realizes, "I can be hurt and feel pain and I am still O.K." Parents can't make their child's pain go away, much as they would like to. Instead, by example, they can convey to him how to respond to it. Ideally, parents should not be trying to shield their child from reality, but instead teaching him to accept and manage it.

When a child *does* express a desire for affection, there should always be an adult available to provide this. When the adult does so, the message that is given is: "Love is abundant, and is there for you when you want or need it." Such a cared-for child is unlikely to become either clingy or distant as an adult. When he or she knows that affection is not scarce and need not be grasped or kept at a distance, he or she is likely to express and receive affection in a healthy, balanced fashion.

Parents will at times find themselves reluctant or unable to give their children this kind of unconditional affection. Unconsciously, they are aware of the depth of their own wounds due to the lack of appropriate affection they received in childhood, and giving this affection to their children instantaneously draws remembrance of this void to the surface of awareness. However, the parent who is willing and able to face this pain without avoiding his or her child's needs for affection stands to gain a great deal. In the words of one family counselor:

> *Allowing your child the expression of true touch from birth can provide a profound healing for the parent. Providing this for your child will elicit severe pain and crises in the adult who has never felt that before. If you as a parent are willing to face and acknowledge this, your child in his or her own inimitable way will draw you out of your own underworld into relationship with him or her.*

One child I know who has learned to give and receive affection freely recently turned to me with her arms open wide and a big smile on her face and said, "You know, I have so *many, many* friends!" It was obvious that from the inside out, this six-year-old child felt loved.

WITHHOLDING AFFECTION

The main argument against providing children with ample, consistent and ongoing affection is a concern about "spoiling them." Ironically, our present approach to raising children, designed to prepare them to face a "reality" that is in desperate need of change, is perpetuating the very "spoiling" it is attempting to prevent. There is a difference between spoiling a child and providing for him or her fully. We literally spoil a child—ignoring her ripening development while her innocence rots away—when we refuse to pick her up when she lies in her crib crying, when we insist that she say "please" and "thank-you" when she is four years old, when we place her in a daycare center to be raised, when we deny her the right to nurse.

Withholding affection from a child should never be used as a form of punishment or reprimand for the child, no matter what the child has said or done. This is one of the most important principles in providing the child with a "hands-on" education concerning the meaning of healthy touch. Conversely, the child should not be given excessive affection for having done something that is good or pleasing to the parent. Clearly, touch communicates on a level that exceeds the intellect. A child can understand when a parent says, "When you hit me that hurt." If his mother puts him down and walks away after a child has hit her, he is devastated far beyond the message that he has hurt his mother.

For parents who regularly show their love to their children through hugging them and other forms of affection, withholding that affection communicates clearly to the child that he is not loved. Of course, the parent does love the child, yet in the moment of rage, when the child is stubbornly acting out an uncontrollable tantrum, the last thing a parent may be feeling is love. Still, the parent does have a developed intellect and she is responsible to her knowledge. She is capable of understanding that while she may be feeling enormously frustrated at her child, that the simultaneous underlying and far more important reality is that she loves him. Therefore, no matter what it takes to do so, she must turn to the child and hold him and love him, communicating through the child's skin that no matter what he did, that he has not lost love.

To parents who have not matured, the paradox that the parent should never withhold affection for any reason from the child, but that the child may withhold affection from the parent at any time and for any reason, may be disconcerting. However, whereas the parent and child are essentially totally equal, they are each operating according to a different set of rules. The child is innocent and the parent is not. This is not true in every case, but it is a valuable meter by which to gauge one's actions in terms of affection with one's child.

The child is innocent because he or she is a child. Although the child will lose this innocence as she grows up, becoming manipulative, controlling, aggressive or withholding at times, any and all of this behavior is learned. If we wish to sustain the innocence in children, we must treat them according to who they are essentially and not who society is teaching them to become. On the flip side, the parent is not innocent. The parent has had ample time to learn to control and manipulate circumstances to get what he or she wants (and children are only too ready to comply in order to secure their primary source of love). The parent knows (if only unconsciously) what he is doing when he refuses love to his child, or when he demands a show of affection in spite of the child's wishes. The parent knows when he is putting himself first, and has the cognitive capacity to chose to act otherwise.

AFFECTION ON DEMAND

If a child blatantly refuses affection, it is not her "natural disposition." However, as children mature, they will desire varying amounts of affection, as well as affection from some sources and not others. (At a certain point, parents will longingly wish that their children still wanted the attention and affection they craved in infancy!) *Affection should always be given on the child's terms.* There's no way around it—it is demeaning and patronizing for a parent to insist that a child give him or her a kiss goodnight, or for an aunt or uncle to demand that the child receive a hug from him or her in spite of the child's obvious displeasure. It is worthwhile and quite humbling to reflect upon how unwilling we are to allow children to decide when they wish to be affectionate.

Even though I consider myself to be open-minded and sensitive to children, I was surprised in visiting an alternative school to find how often I would reach out toward children or stroke their hair when this clearly wasn't their preference. As these children had not been trained in the dogma of, "A good girl or boy always gives Mommy's friend a hug," they would simply ignore me or push me away if they didn't want me to touch them, and I would be left to consider my motivations for seeking their affection. I saw within myself, as well as in many others, that much of our affection toward children is an effort to fill our own emptiness, and to appease our own insecurities about being loved by them.

In our culture, the more common, underlying (and unconscious) motivation for having children is to fulfill the parent's desire to raise children, create a family, receive love, find security, etc. It follows that the unspoken rule would be that children exist to give parents affection, to make them feel good. Ideally, however, parents would have children because they felt that they had something valuable to offer and to share with them. In this scenario, affection would be based—minute to minute—upon the observation of what would best serve the child. Yet we do not live this ideal. Like so many children raised in contemporary mainstream culture, I was taught to give affection on demand. "Come sit on Daddy's lap," "Give your uncle a big kiss on the cheek," "Go give your grandma a hug." Sometimes I wanted to. When that was the case, I had usually already been affectionate out of my own desire. I also remember not wanting to, and the anger and resentment that would ensue as a result—either on the part of the person to whom the hug or kiss was denied, or at myself if I had been affectionate against my will. Already at the age of four I was clearly aware of the fact that I was, at many times, being subtly used for my affection, and I knew well the difference between that kind of touch and the touch I so badly craved.

I was settling down at a restaurant early one evening with my friend Shawna and her young daughter Jenny, when a stranger in a business suit walked over, grabbed Jenny's cheeks, and in a babyish voice said to her, "Aren't you just the sweetest little thing! Look at those chubby cheeks of yours, you're such a darling little sweetheart!" He left giving her a pat on the head. Jenny, who was bonded and knew one kind of touch from the other, turned to her mother and said, "I don't like that man. I don't like that man. I don't want to stay here." We left. As soon as we got out the door, Jenny began to cry. Shawna held her and let her know that she understood her pain.

Some people will read this example and think, "What's the big deal?" "Why did her mother let her get away with that and ruin the whole dinner for everybody?" "Everybody does that with children." That's the problem. That is the way that children are commonly handled, and most have adapted to it. A child who had not bonded and was not taught her own boundaries would never have said anything in response to that man's unrequested intrusion. The ordinary child would have assumed that it was her duty to accept the way the businessman treated her, and say nothing, believing that she did not have a choice about who touched her and when. Perhaps she would even feel good about acquiescing to his intrusion, as the stranger's compliments might make her parents proud of her.

The bonded child, on the other hand, who is given affection according to her own wishes and not demanded to provide it by her parents and others, knows the difference between when touch is needed and when it is excessive. The bonded child can take care of herself, and in doing so helps her caretakers discriminate between a neurotic motivation to demand affection, and a genuine wish to be affectionate with children.

THE NEED FOR TOUCH CHANGES

As the child matures into his or her late childhood and teenage years, her needs for touch and affection change dramatically. As the years pass, though parent and child may continue to have a physically affectionate relationship, the parent's touch becomes more subtle. Particularly when the child no longer wishes to be physically touched by the parent, the parent's touch comes in the form of support and guidance. The parent "touches" the child by allowing her to become who she wants to be. The parent follows her child's unfolding and helps to keep her on track—but only on the course that she has chosen. At this stage in the child's development, the form of the parent's touch may be a letting go—allowing the child to refuse the parent's touch or allowing the child to make her own decisions and learn from her own mistakes.

Flying home from France last summer, I read a magazine article written by the father of an Olympic gymnast. He told the story of how as a child he had been unable to fulfill his own dreams of becoming a professional gymnast. When he met his wife, he told her that his dream was to have a child whom they would raise to be a champion gymnast. The mother agreed, and when their daughter was three-and-one-half years old they brought her to a world famous coach to enroll her in training. Like so many other children, this young woman fulfilled her parents' dreams with precision. She knew instinctually, as well as from her parents' direct messages, that in order to receive their love she had to become a world class gymnast.

What I found most striking about this story was that the parents spoke of their plan to live out their own dreams through their child (even before she was born) as though there was nothing wrong with doing so. At no point in their process was there any consideration about who their child was as a person, what *she* wished to do with her life, and what *she* needed from her parents in order to do so. These parents could have been the most affectionate people in the world, but they were not giving their child affection. Any genuine affection that they may have felt for *her* was secondary to the unfulfilled aspirations they had for themselves.

Though simply stated, these principles are not easily enacted. It is important to reiterate that we are trying to communicate something to our children that we have little or no experience of within ourselves. This is a tall order, calling forth risk, experimentation and creativity. What we supposedly "know" about raising kids is often the very hurdle that we must overcome. We can think or feel anything we want to about our child or about a particular situation, but at the same time, the gesture of touch that we communicate, as well as the intention that must be generated to enact it, is what counts.

FATHER BONDING

> *I knew my son needed me,* began Joe, a tool and dye maker, *and I was going to do anything I had to do to be with him. I used to take off work just to hang out with him. Fortunately, I was a good liar because I used to tell my boss all kinds of things just to spend time with him. I'd strap him on my back and we'd go hiking or walking or out fishing. I wasn't protecting him in the same way as his mom. I was loving him and touching him but I wanted him to know the world...When my wife and I got divorced I quit my job and found a way to be self-employed. He was with me all the time. I taught him to use all the equipment I was working with. He'd help me out when he wanted to, or just hang out. I was lucky though. For one, I knew enough to know that a young boy needs his father's touch, and secondly, I was a good enough worker to be able to make it on my own in business...He is grown now*

and there's a big difference between him and my other son whom I wasn't able to raise and who never had a man to spend time with.

Bonding with the father will provide the child with something they cannot get from the mother. If a child is deeply bonded to the mother, but not the father, he or she may still grow up feeling loved, but will feel one-sided and imbalanced, as though something is missing. Children need a role model of the opposite gender to balance the masculine and feminine energies contained within them. Bonding with the father—not just "the masculine" but with a man—may be the single most important determining factor in that child's relationship both with the masculine aspect of themselves, and with all of the men that they will encounter throughout the rest of their lives. Furthermore, the child's ability to live constructively and sympathetically in a masculine-based society will result from having had a positive reference point for the masculine that is constructive and active without being aggressive.

What takes place during the time of bonding becomes internalized in the child—it becomes a part of their own internal make-up, including their worldview and the way in which their experience is perceived. If father bonding has occurred, the male child is more likely to grow up to be productive, confident, clear and capable of seeking warmth and companionship from other men as well as from women. The female child who has bonded properly with her father will mature into the strength of her womanhood, able to be as strong as she is soft, and will come to relate to men with respect and honor instead of fear or condescension.

Again, it comes down to touch. Children need the loving touch of a man to feel secure, grounded and capable. This rarely happens in a world in which men are required to work eight to ten hours a day, and in which, on a cultural level, men have not been given the support and encouragement to touch their children...but it could.

There are stages of the child's life when they will draw more on a masculine source of support and stages in which they will need the feminine. During the first years of life, the child needs the mother—when possible—like nothing else. Yet when it is time for the child to move out of the mother's protective embrace and into the surrounding world—often occurring around the age of seven—a father or male presence is often needed to help release the safety valve and transition the child into the outer environment. When children have bonded with both mother and father, that bonding will serve as an anchor to support them through the various stage-specific needs that will arise.

Jake, my nine-year-old son, had just moved back from Nevada and was living with me. I had not raised him since he was three. Unfortunately, I had a full-time job, so my wife would pick him up from school and they would spend the afternoons together playing and doing his homework. They had an excellent relationship, but Jake began to have violent mood swings and would often refuse to get out of the car when I dropped him off at school. I asked a trusted friend for advice and he suggested that Jake needed to spend time with a man. He said I should spend an unreasonable amount of time with Jake—as much as possible. I'd get up with him in the morning, take him to school, pick him up from school and be with him in the afternoon while my wife worked my business. He's really taken to it, but it's sure not easy. All I can say is that at a certain stage of his life, a kid needs his father. A father gives a kid a whole different perspective—a kind of inner strength that he doesn't even know he possesses.

Having said all this, the reality of our present culture dictates that there are many single-parent families in which one parent has died or there has been a divorce. There are gay families and growing numbers of people living in alternative settings. The mother cannot be a father, and the father cannot be a mother—and that doesn't have to be a problem. One option in such instances is to enlist the support of a male friend or relative who is willing to spend time with the child, and to serve as a mentor or role-model. This can be a mutually fulfilling situation for both the mentor and the child, and with the right person, can provide the child with the guidance he or she needs as at this time.

THE PROBLEM OF INSTITUTIONALIZED DAYCARE

In spite of what has been said above about the child's need for his own parents affection, at the present time there are more children in full-time daycare than ever before, and this concept of daycare centers is gaining in popularity in many countries—particularly in the Western world. The percentages are rising!

The *American Demographic Magazine* reported that half of all working women who are new mothers, and two-thirds of working women over thirty who have a new baby, are back at work before their infant is one year old. Furthermore, two-thirds of women with preschool children are working mothers. The Bureau of Labor Statistics estimates that by the year 2000, women will account for approximately 14,000 of the expected 21,000 new workers.[1]

In many cases, particularly that of the single parent, this is a difficult predicament. Many single working mothers cannot take off *even one day* of work to look for adequate daycare for their child, and may not be able to afford the better options available when they do find them. But a bigger problem yet is that for many, the question of their child's psychological and emotional needs never even arises when considering placing their child in a daycare situation. The industry of daycare, backed strongly by a non-child-centered set of cultural values, promotes and glorifies the use of this surrogacy. Feminists will often argue that all women—mothers or not—have a God-given right to a full-time job, and that the so-called patriarchal mandate to stay home and take care of her children interferes with this fundamental right. This argument is essentially faulty in that it is based solely on the desires of the mother, and does not take into consideration the needs of the child. Whether it is the mother or the father who will take responsibility for caring for the child is not nearly as crucial an issue as the need for one of the parents to do so. A daycare situation where the child is not only unlikely to get one-on-one support, but will probably wind up in a room of many; and a place where bonding is only available from nine to five, is probably not going to give that child what he or she needs.

Can a child develop a bonded relationship with these part-time daycare "parents" who are often overworked and who, due to the increasing legal dangers of touching young children other than one's own, are unlikely to give them the affection they need? (See "Two Wrongs Don't Make a Right," Chapter Four) And if not, how does a parent balance the needs of the child, the needs of the family, and the needs of the mortgage? The baby is going to attempt to bond with something—and if he can't bond with the mother or father, he will attach himself to a blanket, a thumb, or the lady at the church who smiles at him on Sundays. If parents want their child to bond, rather than to compulsively attach himself to someone or something, it is best that they do the bonding themselves.

Many mothers who return to work soon after childbirth could do otherwise, though they may have to sacrifice the fancier car, the new furniture, the Christmas vacation, etc. But many other parents—particularly single mothers and fathers—*do not have a choice.* What are the options for parents? Some mothers find employment that they can do from home. Others are able to find daycare centers near to their jobs and arrange to take short periods of time to be with their child and tend to any needs that may arise (this is particularly useful for young children who need to nurse). My brother gets up with his child at 5 a.m., spends the morning with her, and takes her to work with him at his carpet warehouse for two hours so that his wife can sleep and tend to the other needs of the household, thus being available for their child throughout the rest of the day and during the night. Small groups of parents arrange their own daycare groups so that they know who is with their children and can have input concerning how their child's day will be spent.

The process of bonding is inherently intelligent. As parents make the gesture to bond with their children, the bonding itself informs the parent of their child's needs, and gives the parents the necessary desire and energy to serve their child.

SEXUALITY IN CHILDREN

Although rarely spoken of, sexuality is a touch-related issue that all parents will confront in raising their child from birth onward. Sexuality does not suddenly "develop" when a child hits age twelve or thirteen—it was there all along.

As will be discussed extensively in Chapter Six, sexuality is much more than sex. It is not something that is exclusive to the realm of adulthood and taboo in children. To children, the vagina or anus is just another hole; the penis is just another thing that sticks out. Sexuality is not a problem for children until we make it into one for them. As a culture, we are afraid of sexuality in children because we are uncomfortable with our own sexuality. If it is considered socially inappropriate for a mother to breast-feed in public—the most natural of events—how will the culture embrace the fact of children playing with their penis or vagina when it is right there and it simply feels nice to them? One father shared:

> *They're going to play with their genitals, and if you're like me and really repressed, you're going to think, "What should I do?" because they really play with them. They don't just touch them. They lie there feeling good. If you get uptight and brush their hand away, you're telling them it is bad. You just have to keep your mouth shut and your anxiety to yourself.*

Working in battered women's homes, private schooling situations, and even walking through grocery stores, I have often seen parents grab their young children's hands away from their genitals, or give them a whack on the back of the hand when they were playing with them. Sexuality only becomes bad, dirty, taboo and inappropriate when we teach children that it is so. Oftentimes, the less said to the child the better. The more attention drawn to it, the greater the energy, fear, interest, apprehension and bias that will be created in the child, instead of allowing sexuality to be a natural thing.

Similarly, sexual play among children is natural. If parents are attentive to their children, they will know what they are up to. Wise parents will set boundaries—for example, no placing of objects inside of themselves or anybody else (because it is dangerous, not because it is bad), or no playing with themselves in public or when adult members of the opposite sex are around who are not their parents (only because in our culture there is an already-present state of sexual confusion and misconduct. In many other cultures this would be no problem). One man shared:

> *When I was nine and had just learned about kissing, I took my sister into my room and told her I wanted to show here something. I kissed her on the lips and it was really fun. But soon after that, I started to get the idea about the incest taboo, though I didn't have the words for it, and for years afterward I felt guilty about it, thinking I may have had incest with her. What wasted guilt. It was spontaneous, natural and healthy child's play—that's all!*

When children are allowed to be naked in their homes, and to see their parents naked, they come to understand nudity as natural and normal. On the other hand, when a two-year-old is constantly being made to cover her genitals, and when her parents are uncomfortable with the child's nudity, she forms an assumption that there is something that must be hidden. When young children are left to have their nudity as they please, as they grow up they naturally gravitate toward the wish to wear clothes in order to "fit in" and to integrate with the surrounding culture.

Children can be answered honestly and directly when they ask about sexuality—they are more capable of such understanding than we commonly imagine. Parents may choose to answer their children in language that is simple and straightforward, rather than confusing their already-alert minds with stories of birds and bees and storks—which is demeaning to their intelligence. An honest approach is best. Granted, many parents may feel uncomfortable in speaking about sexuality—this is understandable given that most of us have been raised in cultures that do not speak openly about sexuality. Such discomfort is normal because of our training, but despite this uneasiness, parents are learning to speak clearly and simply to their children, answering what is asked in a language that they can understand.

In most traditional cultures, there were rituals, ceremonies and stories that would be told in order to teach the child about sexuality, not leaving the burden on the child's parents themselves. But we do not have that luxury. If we want to change our culture from one that has a largely aberrant sexuality, to a culture with a healthy sexuality, we can begin to catalyze such change by developing a healthy approach toward discussing sexuality with our children. And in order to do that, we need to set our discomfort aside, and be willing to answer our children's questions honestly and clearly.

Children benefit from knowing not only about sex, but about masturbation, menstruation and so forth. I know a young girl who for a time was fascinated by menstruation. She would often wear her mother's sanitary pads in her underwear, or play with tampons. After awhile, the fascination wore off. When sexuality is not repressed in children, they will naturally cycle through the various stages of sexual development at their own pace.

When I initially began to travel in Third World countries, I couldn't imagine how parents were producing all of these children when the whole family slept in one room. This question was based on the assumption that sex could only occur in a bedroom with a locked door, or in other private quarters. Radical as it may sound, when parents share a bed with their young child (See "Bedtime," Chapter Two), sex between parents happens right there in the bed while the child is sleeping and is not in need of the parents' attention. Or, parents sometimes place their sleeping child on a quilt on the floor, or move onto the floor themselves. Should the child wake up and need something from the parents, the child's needs are placed first. When sex is simply occurring as one natural element in the child's entire field of life experience, even before he or she understands what it is, it becomes integrated in him or her as just that—natural, normal, and no big deal! Children develop their sexuality and their ideas about sex from their role models. Whether parents hide it behind locked doors and don't speak a word about it, or whether sexuality is openly acknowledged in the household, children pick up on it. They know something is going on, so to keep it hidden from them is to convey a very specific message to them—that sex is secretive, taboo, and not to be discussed.

In a sexually dysfunctional culture, some parents will probably experience some degree of sexual feeling toward their children at various times. Whether parents admit to it or not, most will probably find their young child's body to be sensual and beautiful to watch—that's because it is! It amazes me that this appreciation is such an unspoken reality, and it can only be attributed to the major confusion prevalent in our society concerning sexuality. Since these sensual feelings are taboo, when they arise in parents they are immediately suppressed, put out of the mind, denied, and often result in the parent feeling guilt or shame. It is often not understood that these feelings are commonly indicative of a desire for contact and intimacy as opposed to a wish to have a sexual relationship. In a touch-starved nation, where sexual dysfunction runs rampant, many people cannot make the distinction between non-sexual intimacy and sex.

If a mother cannot acknowledge her "sensual" feelings toward her young son, and understand them for what they are, she is far more likely to subtly treat him as a lover or surrogate husband. She will probably not take him to bed with her or expect him to take her out to dinner, but the covert ways in which she expresses these feelings has left many a man confused about his sexuality in later life, feeling that he is betraying his mother by being with "another woman" when he gets into relationship, or feeling guilty for feeling angry at her "for no good reason," as she was always "a good mother." Similarly, the unacknowledged sexual feelings of a father toward his daughter (or son) is one factor that may lead a father to have an incestuous relationship with his child.

At first, sexual or sensual feelings towards one's children may be hard for parents' to admit, but such feelings are not uncommon in a sexually dysfunctional culture. Parents address this by beginning to clarify and understand these feelings in healthy ways, within themselves, instead of acting them out with their children, and thereby save their children many subsequent years of confusion and difficulty in their own intimate relationships.

Because some parents feel these feelings does not mean they should not touch their children, or that their touch is sexual. Parents who are honest and clear with these feelings, and willing to take responsibility for them, are the least likely to abuse their children. Fathers at the present time are often left in a difficult bind—if they don't touch their children they may be considered distant, and if they do touch them they may be suspect of abuse. It is an unfortunate situation, but when the father is able to examine himself honestly—free of his need to deny and repress the uncomfortable aspects of his own inner world—he knows exactly what is appropriate and what is not. The same is true of the mother. They simply need to act on what they know.

A friend of mine recently told me that one night after dinner her three-year-old daughter jumped up onto Daddy's lap and asked him to play with her genitals. The father responded calmly and lovingly that this was not how their relationship should be, and that later in life there would be a man who would fulfill this role. The child was not shamed, blamed, or criticized—just guided in the right direction.

On the flip side, a client of mine who came into therapy because of difficulties in her sex life with her husband later reported memories of standing in front of the mirror as an eight-year-old and hitting her vulva saying, "Bad. Bad." Already at eight years old her future had been determined—she had no hope of having a healthy adult sex life unless she were to do extensive healing work as an adult.

When childhood sexuality is denied, misunderstood or abused, and when children are not taught about their sexuality, they often enter into a period of excessive experimentation in their early teenage years. Teenage pregnancies are the sorry consequence—resulting either in far too many mothers and fathers who are totally unprepared to be parents, or in abortions that are often traumatic and could otherwise have been avoided.

At the abortion clinic where I worked in my early twenties, I saw many women far younger than myself come into the clinic on the weekend, or even skip out of high school during the week, to get an abortion. Usually they would arrive with a girlfriend, rarely with parents or boyfriends, who often didn't even know about the pregnancy.

I remember one particular incident in which a seventeen-year-old young woman came in alone. She was perfectly groomed with make-up, stylish clothes, painted nails and a new haircut, and told us she was going to cheerleading camp the following day. When I asked her how she felt about the abortion, she said, "Oh, just fine. It's no big deal to me!" Her cheerfulness and apparent lack of feeling had me suspicious and attentive, but she continued to chat away with me while we went into the operating room. As soon as the abortion began, she fainted, and we had to bring her to the hospital when she did not wake up after an hour. By going unconscious, her body paid the toll for the repressed anxiety that she was unable to feel.

When children are taught the beauty and naturalness of sexuality and of right relationship to sex from the very beginning, they grow up inherently understanding the difference between the two, and are less likely to take on the distorted attitudes of the surrounding society.

The gap between what needs to happen in our world in terms of touch, and the reality of where we are and the direction we are going should be increasingly apparent. A culture of unbonded children will grow up to become a culture of unbonded adults, and it is the unbonded adult who is the most likely candidate to become violent and abusive. Abuse—the greatest misuse of the power of touch—is an often misunderstood phenomenon in our culture. Whereas obvious manifestations of abuse are relatively easy to identify, less apparent are the subtle and invisible causes of such behavior. Chapter Four discusses the relationship between touch-starvation and what I refer to as a "culture of abuse," for it is only in the understanding of the roots of violence and abuse that we have any hope of remedying the present situation.

CHAPTER 4

The Wrong Kind of Touch: A Culture of Abuse

Aggression is not strength. It is exactly the opposite...The advocates of aggression...say: "Life has been hard for me. I've been knocked around, and it's made me what I am. Let it be the same for my children." Really what they are saying, without admitting it, is, "I've suffered. Why shouldn't others suffer too?"[1]

– Frederick Leboyer

I was eight years old and on a camping trip with my parents. My older brothers and I were happily playing with some new friends we had met from the next trailer, when suddenly their mother's voice came roaring toward us, "Jeerremiee David Hopkins, get your goddamn ass over here right now before I drag you over here myself!" Immediately the boy's face went pale, his jaw started trembling and we were all silenced. I remember thinking to myself, "What could he possibly have done that was so terrible?" Jeremy was too terrified to budge. Sure enough, his mother came over, grabbed him by the collar, and dragged him back to their trailer. Apparently, eleven-year-old Jeremy had not washed the dishes from breakfast as he had promised to and now it was lunchtime and the dishes were still dirty. We all watched through the trees as his mother, still screaming at the top of her lungs about what a "good-for-nothing, irresponsible son" Jeremy was, took a huge wooden spoon and raised it above Jeremy's behind. I'll never forget the look on his face—sheer terror, like the kind you see in the movies when somebody has a gun held to their head. Both my older brother and I were crying in anticipation. The smack on the butt she gave him with that spoon wasn't so hard, but Jeremy was already traumatized. Then she laughed. "Ha! I hardly have to touch him and he's already crying." In that moment, all of us present hated that woman. In fact, beneath her own rage, she probably hated herself.

This type of scenario—an outright and unabashed abuse of the gift of touch—happens thousands and thousands of times a day throughout the world. In fact, in many places it is not even considered a problem. But the frequency of its occurrence does *nothing* to diminish the lasting damage that it has on the child's soul—the child who will grow up to be an adult who does the very same things that they once hated their parents for.

This chapter explores the issue of abuse on all levels—from abuse done by individuals to abuse enacted by the culture—so that the reader may understand what he or she is faced with in living in a touch-starved nation, and what the consequences are of our failure to touch and love one another appropriately.

As profound as the capacity is for loving touch to bring us closer to ourselves and to allow us to feel loved by and connected with others, such is the capacity for abusive touch to devastate the soul and alienate us from the world, resulting in feelings of despair and worthlessness. The power of the human touch—by birth contained within each of us—must be used, or not used, with intense sensitivity and consideration. Many people fail to recognize that their choice to touch or not touch their children, intimate partners and friends, as well as *how* they choose to touch or not touch them, is the difference between creating either an environment of love, or one of unlove.

To take an objective inventory of the state of abuse into which we as a culture have "evolved" can be extremely unpleasant, if not altogether heartbreaking. Yet in order to deeply understand the power of touch and the necessity for it, and to seriously intend to bring loving touch into our lives, we must be willing to look straight in the eye of what happens when this power is wielded unwisely, as is so often the case.

What is most disheartening is that abusive attitudes and behaviors have now penetrated our culture so deeply that they have become a cultural "norm." It is so common for adults to neglect their children, spank them, scream at one another, threaten and hit each other, as well as to abuse one another in a myriad of subtle, psychological ways, that many people don't think twice about it. Because this abuse has been "taught" to us by our parents, and was taught to them by their parents, and is enacted by so many people in the surrounding society, many do not stop to question if such ways of relating to one another are sane. Instead, a low-grade abusiveness becomes part of the larger cultural behavior, escaping our recognition of it as such.

Recently, a close friend of mine was the victim of a violent crime, the kind you read about on the front page of the paper—rape at gun point, kidnapping, robbery. She barely escaped with her life. This was my friend—the one I drank coffee and ate pancakes with on Sunday mornings; the woman I could call at any time of the day or night when I needed somebody to talk to. Violent crime was suddenly a stark reality for me. It wasn't something that happened to "those people out there." Still, the majority of abuse happens much closer to home—in the home. It happens in shockingly obvious forms as well as almost undetectable but equally damaging ways. The damage created by abuse cannot be measured by the type of abusive activity enacted, but by its impact on the individual who has been abused.

The purpose of exploring abuse here is not to indict or blame any person or persons, but to further our understanding of ourselves and our world. Therefore, defensive statements such as, "I only hit her once," or, "Yelling at somebody isn't *really* abuse," are irrelevant to this discussion. Abuse is in the body of the beholder. If we as a culture really *felt* how others are affected by our own abusive behaviors (felt it in our guts and in our broken hearts), abuse could never occur on the level, and to the degree that it does at the present.

WHAT'S ALL THE HYPE ABOUT?

The US Advisory Board on Child Abuse and Neglect documented some 2,000 annual violent deaths of children four years old and younger due to child abuse and neglect. Another 18,000 have been permanently disabled, and 142,000 seriously injured due to violence.[2] - Gerald Kreyche

Nobody lives in the 90's and escapes confrontation with the issue of abuse—and it is not going out of style. Whether we are reacting to our own abuse, or responding to those who are reacting, we are all in a state of alert with no neutral or objective party to clarify what is what. What is all the drama about? What is agitating us so?

People tend to be divided into four different camps around the abuse issue. The first group is comprised of those who recognize that because of the cultural zoom lens focusing in on abuse, that they have finally found the long-awaited forum in which to speak out about the violence they have been enduring for years. Some may currently be experiencing abuse, others may be struggling with the long-term effects of abuse, whereas for another group this may be the first time they have heard a name for what has been driving them crazy for decades. People who have been abused often have lost their voice—they have lost the courage to speak out about what has happened to them and to stand strong within themselves. For centuries, women who have spoken out about abuse have been ridiculed or publicly castigated, and children have been disbelieved and often subjected to further abuse as punishment for having spoken up. Abuse has been kept secret, mislabeled (e.g., "A little spanking from time to time keeps children in line") or altogether repressed. The past decade has marked the first time in the history of many countries where there is a forum in which victims of abuse can speak up and be heard to any degree.[3] They have, at long last, the possibility of connecting with others—in support groups or 12-step groups—who understand what they have gone through. There are hotlines to call and people with understanding to turn to who won't disregard their claims.

Society tends to react defensively to the voices of these victims because what it often hears is an angry accusation that feels blaming and unjust. That's understandable—by the time most people who have been abused realize what has actually happened to them and those around them, they *are* mad. A young man who repressed memories of his third grade gym teacher making him masturbate in front of him is going to be angry when he remembers what happened. A woman who has been in two marriages in which she was beaten by her husband, only to finally recognize that she has repeatedly gotten into this situation as a result of having watched her mother be abused by her father for the first two decades of her life, is going to be angry. A middle-aged man who, as a boy, laid quietly in bed unable to sleep because of the sound of his father hitting his mother in the next room, believing himself unaffected because "it didn't happen to me," is going to be pissed off when he understands this to be the root of his lifelong difficulty with intimate relationships.

Unfortunately, in the midst of expressing their anger, these victims of abuse may neglect to consider into whose ears they are screaming, or to take the time to communicate in a way that others are able to hear and relate to. They have finally broken their silence and they have a lot to say. They want abuse to stop *now,* but they don't know how to stop it. Many do not understand the psychology of abuse, nor the politics of it, and thus their outcry feels insulting to the general public, which dismisses them as angry feminists or wimpy men. Yet, even those who have a precise understanding of abuse, and who *do* know how to speak about it in a clear, direct and simple fashion, are also met with rebuttal and disregard. People don't want to hear about abuse. It is fine as a nighttime news story or in a soap opera, but when they are asked to examine how this shows up in their own homes and in their own lives, they immediately look the other way.

The second camp of individuals responding to the issue of abuse is made up of both those who have abused others, as well as those who have been abused themselves, and are desperately trying to keep the memory at bay. They are the ones who say that the whole issue is over-dramatized, blown out of proportion, and made up by children who are prone toward lies and exaggeration. To be discussed further in the section entitled *A Culture of Denial*, these people feel extremely disturbed by all the attention the issue of abuse is getting, often not even knowing why this angers them so. Parents who thought they were raising their children normally, but who did not hesitate to "discipline" them by screaming and hitting them, or having them wash their mouths out with soap, etc., are suddenly hearing this behavior labeled as abuse. Those who have justified their abusive behaviors by saying it is no worse than what their parents did to them, or who quietly dismissed it by telling themselves that these were private affairs that would not leave the home, are now finding their deeds written up in newspapers and spoken of in books. There are also those who have been abused by parents, spouses or strangers, and who have tried mightily to forget or ignore it, unwilling to face the fact that the people who supposedly loved them would hurt them, and unwilling to acknowledge the reality of their underlying pain. They do not want to hear abuse spoken of because it stirs their memory. Each article they read, each story they hear, increasingly threatens to uncover the wound that they have tried for so long to forget.

A third camp of people consists of those who capitalize on the first two. They are reaping the profits of the abuse market. Abuse has become a wise investment for entrepreneurs, journalists and screen-writers alike. The public has always been drawn to the sensational aspect of abuse—

preferring to watch movies and to read news stories that depict violence rather than honest love. Rage, injustice and mistreatment are portrayed on television more widely than any other subject, and there are now hundreds of books available on the "abuse self-help" market—whether your forte is sexual abuse, verbal abuse or violence against animals. The psychotherapy market is booming because so many people are psychologically wounded, and journalists jump on the latest story of a movie star revealing that she was repeatedly raped in her father's home. Dishonest people are seeing that they can file a multi-thousand-dollar lawsuit, saying that their child's teacher touched him inappropriately, or that their boss approached them with sexual propositions at work. Abuse is hot, and the go-getters are onto it.

Lastly, the fourth camp of people is comprised of the masses who could simply not care less about the issue of abuse altogether. They are far more interested in finding out about the latest NBA scores, or reading a Harlequin romance novel, or going shopping at the mall, than they are in even considering the fact of the epidemic of violence. In some of their homes, although abuse is spreading like the plague, they create the appearance of things to look like the "Brady Bunch." They simply don't want to hear, think, or speak a word about abuse. The technical term for this is *denial,* but who can blame them? In some way, ignorance is bliss. Ignorance is not reality, but reality is a tall order for many people.

Given that there is so much talk of it in the media, so much denial about it in our culture, and so many misconceptions and confusion about what actually constitutes abuse, it is useful to draw some distinctions about the various forms of abuse before proceeding into a more in-depth discussion of its effects on a touch-starved culture.

WHAT IS ABUSE?

According to Webster's College Dictionary, abuse is defined as follows: *to treat in a harmful or injurious way....to speak insultingly or harshly to or about....to commit sexual assault upon....harshly or coarsely insulting language....bad or improper treatment....rape or sexual assault.*

Abuse is a very tricky term to apply. On one end of the spectrum, executives who make seductive comments to their secretaries are being taken to court on the grounds of abuse; on the other end, people are getting away with literal murder on the grounds that they were temporarily "insane," (which implies that we live in a culture in which the norm is sanity). Nonetheless, it is important for us to continue to refine our understanding as to what constitutes abuse.

If people are honest with themselves (though most abusive people are not), they know when they have harmed another person. How an individual life is affected by abuse is really the issue. One parent might regularly yell at his child and whip him with a belt and that child may manage to grow up fairly well adjusted—at least superficially; whereas another child may be broken in two as a result of this same type of treatment, and pay for it throughout his adult life.

Sexual Abuse

Sexual abuse is probably the most insidious misuse of the power to touch that human beings inflict upon one another. It is comparable to murder, only it is the soul of the individual that is often killed and not the body. It may occur between parents and children, spouses, boyfriends and girlfriends, teachers and students, and strangers. The National Center for Child Abuse and Neglect defines sexual abuse as, "Contacts or interactions between a child and an adult when the child is being used for the sexual stimulation of that adult or of another person. Sexual abuse is any inappropriate suggestion or actual sexual exposing or touching between adult and child."[4] When sexual abuse occurs between members of the same family it is called *incest*.

When a mother fondles her eight-year-old son's genitals, it is incest. When a young woman refuses to have sex with a man whom she has spent the evening with, and he insists regardless, holding down her arms, it is sexual abuse. When a father gets in bed with his daughter every night to "tuck her in," but instead rubs his aroused body against hers, it is sexual abuse.

There are also other forms of sexual abuse that are much subtler but may result in similar forms of damage in the psyche. For example, the male child of a divorced woman who is treated by his mother as a surrogate husband will inevitably wind up confused about his sexuality. The young girl who is "raped" by her third-grade teacher's lustful eyes, and then called flirtatious when she speaks about this, will suffer at least some of the effects of sexual abuse. Although a brilliant psychoanalyst, Sigmund Freud did a great public disservice when he overemphasized the child's sexual desire for his or her parent, but did not speak to the parent's desire for the child. As discussed in the section entitled *Sexuality in Children* in Chapter Three, in a sexually dysfunctional culture, it is fairly common for children to have sensual or erotic feelings directed toward the parent of the same, or the opposite gender. It is also common for some parents to have erotic or sensual feelings toward their maturing children. However, the responsibility rests entirely on the part of the parent to refrain from responding to their child's feelings or seductions and from acting on their own impulses—both literally, as well as in subtle ways.

Physical Abuse

A mother takes her son over her knee and spanks him with her hand.

A teacher whacks her student's fingers with a ruler.

As his wife runs out of the house to the car after a fight, her husband chases her and grabbing her arms, pins her against the car until she gives him the keys.

Physical abuse is touch backed by aggression and a desire to inflict pain—whether it be a so-called disciplinary measure, or an outright violent attack. Yet another gross form of the misuse of touch, it is a painful reminder of how little responsibility human beings are taking for their intelligence. Physical abuse is the easiest form of abuse to recognize, and it is what most people think of when they hear the term "abuse." However, one needn't wind up with a black eye or a broken rib in order for a physical attack to constitute abuse. A slap across the face, chasing another person and threatening them, pinching someone until they cry, tickling someone until they are struggling to get away, digging one's nails into another person's skin—all are forms of physical abuse.

Verbal/Emotional Abuse

Verbal and emotional attacks are abuse. A man who degrades his wife in front of their friends is verbally and emotionally abusing her. A mother who returns home from work and upon seeing that her son has not cleaned up after himself screams at the top of her lungs at him, "You good-for-nothing slob. Go to your room and I don't want to see your ugly face again for the rest of the night," has literally taken her child's face and rubbed it in the dirt. Similarly, when a child watches her sibling being degraded and torn apart by their parent, that child suffers the abuse. A client who had been verbally abused continually throughout her childhood by her father, and watched her mother and siblings undergo the same, expressed the sentiment of countless other children in the following:

> *I always wished he would have hit me and scarred me and made me bleed so that he could have seen what he was doing to me. He often threatened to and I would turn right back to him and say, "Do it. Then maybe you'll be forced to see what you're doing to me all the time." But he never would. I actually used to pray that he would hit me so that others could see what was happening and maybe they would help. I was never afraid. No wound on the outside could be more painful than the ones I was already enduring on the inside.*

Neglect

Neglect is the failure to support another individual emotionally and to tend to their needs, often including the refusal to give nurturing physical touch. Children who from the age of six come home from school to an empty house and have no further human contact until evening, and whose baby-sitters are televisions and video games, will suffer neglect. The child whose parents are so absorbed in one another that they ignore their child, or who give a lot of touch to their spouse and little to their child, will suffer neglect. The child who is never held by his or her parents, because the parent finds it awkward or embarrassing, will suffer neglect. The consequences of neglect are the failure to bond and the subsequent difficulty in forming intimate relationships in adulthood. Neglected individuals grow up feeling unwanted by the world and often repulsed by touch and affection. They become hardened, and may harbor deep feelings of rage and resentment at their parents or the world. They develop an attitude of, "I don't care about anyone or anything," in order to disguise a much deeper pain about feeling unloved. Lee Lozowick explains:

> *Neglect is not just to throw your baby in the crib and walk out the door (to the bedroom). Leaving your child when he does not want to be left—that's neglect, too.*[5]

Subtle Forms of Abuse

There are numerous forms of subtle abuse. When a jealous husband does not permit his wife to talk to any other men—it is abuse. When a teacher shames a student for giving an incorrect answer and tells him she is doing it for his own good—she is abusing him. When a mother fails to defend her child when she knows her husband is abusing the child—it is abuse. And the issue becomes still subtler. A father who is exhausted and wishes for time alone confuses his wide-eyed and alert child by telling her that *she* is very tired and that is why he is putting her down for a nap. A boss who, frustrated by his difficult relationship with his wife, acts this out by being controlling and inconsiderate toward his employees, may be unknowingly creating an abusive work environment.

Parents who point out everything that their child does wrong, but fail to acknowledge his or her gifts, skills and achievements, however small, are contributing to an inner environment of abuse within the child. Subtler still are the countless ways that we fail to respond to the needs of our spouses, children and friends because we are too absorbed in our own experience to notice what is happening for them, and the endless emotional "jabs" we give one another throughout the day in our wish to put another person down in order to gain the upper hand. Abuse is going on continually. The rare moments we are able to glimpse the pain that we cause others throughout the day reveal the bittersweet reality both of how easy it is to hurt another person, as well as the possibility of what it would mean to genuinely love one another.

Self-Abuse

I often ask clients, "How would you feel if your best friend talked to you the way you talk to yourself?" The response is usually one of shock and horror. For many people, their entire day consists of an endless dialogue of criticism and guilt that may go something like this:

> ***When am I going to start exercising before work? Why did I drink two cups of coffee again when I am trying to cut back? Why do I let this traffic get to me so much? How did I spend the whole morning at the desk and get nothing done? I shouldn't have eaten so much at lunch. I wish I dressed like so-and-so. Will I ever get a better job? Why did I marry a woman who was going to spend my whole paycheck shopping? I wish I would stop spending the evening in front of the television set and get outside more. My***

wife isn't attracted to me. Why can't I ever get everything done in time to get a full night's sleep?

On and on this goes, throughout the dream state and into the next day. Many people are steeped in a climate of internal abuse, irrespective of their interactions with anybody outside of themselves. This dialogue is so incessant and has been going on so long that it is almost unnoticeable to the individual whose reality has become one of intense suffering. The person is convinced of their worthlessness and overall "badness." Underneath all the abuse they are really crying out, "What's wrong with me? If only I was perfect, or even better, I would be loved." And still more deeply hidden is the core pain of, "Why didn't they (my parents) love me?"

A climate of internal abuse will inevitably result in difficulties within all of one's relationships. There is a continual expectation that others hold the same judgments and criticisms as the individual has about himself or herself, leaving the individual afraid and constantly on the defense. This fear of abuse emanates from people and draws this reality to them, thus affirming their beliefs. For example, the woman who was abused as a child and is now terrified of being rejected and unloved may isolate herself as an adult, apprehensive of extending herself to others. When others don't respond to her, her belief in her overall "unlovablility" is affirmed.

Other forms of self-abuse include alcohol abuse, drug addiction, eating disorders, self-mutilating behaviors and repeatedly placing oneself in abusive relationships, friendships, work environments. Although in part designed to ward off hurtful feelings, these are all ways in which the already-hurting people inflict further pain onto themselves.

TWO WRONGS DON'T MAKE A RIGHT

In September 1996, six-year-old Jonathan Prevette, a first grade student in Southwest Elementary in Lexington, North Carolina, was accused of sexual harassment by a school bureaucrat, and barred from his class for a day. Jonathan had planted a kiss on his classmate's cheek. (His parents were offered $100,000 for movie rights.)

A "touch paranoia" has swept the nation. As a result of so much attention on the issue of abuse, combined with the multi-million dollar lawsuits that have followed, as well as people's insecurities concerning how to touch one another appropriately, many people have responded with a concrete decision to stop touching one another altogether in particular environments in order to prevent being misunderstood, as well as to protect their savings account should they be accused of touching another person inappropriately.

Unfortunately, two wrongs do not make a right, and individual decisions as well as laws that mandate groups of professionals to refrain from touching one another are going to have a boomerang effect, returning with an exponential blow to the individuals as well as the society which has taken this initiative. It is the individual who ultimately suffers the isolation and loneliness of a personal life or work life characterized by a lack of intimacy, and it is the society which has outlawed love and disregarded the basic goodness of its members that will be responsible for creating an increasingly impersonal world in which genuine affection is a precious commodity with a price tag attached to it.

A man who worked for Boy Scouts of America for many years stopped his work with them when, after several accounts of abuse within the Boy Scouts organization had gained national attention, a rule was instated that a male Boy Scout leader was not allowed to be alone with a young boy—that there must be two or more boys present for any interaction to occur. This man said that the great value he had found in the Boy Scout movement was that young boys who were having difficulty in their lives at home and who were unable to talk to their fathers, or who perhaps didn't even have fathers, had the opportunity to form a relationship with, and to confide in, another mature man, and that this intimacy was not possible to the same degree when two or more boys were present.

The schools are a primary target for anti-touch laws. More and more states are making it illegal for teachers to touch their students, and even in those states that have not outlawed touch in schools, it is increasingly risky for teachers to touch their students. The consequences of these laws alone will be devastating to thousands and thousands of young children across the country. How is it that a teacher who hugs a crying student can be suspect by law for touching him inappropriately, yet that very same legal system allows her to whack that student with a paddle?[6]

Many children do not receive affection at home—period. Abuse and neglect rates are high—much higher than most people dare to imagine. A caring teacher who is responsive to, and respectful of the child as an individual may be the sole source of genuine affection the child ever receives, as well as the only role model of healthy touch he or she is exposed to. If a child is being hit or neglected at home, and treated as though he has no human needs for affection by his teachers at school, how could he be expected to grow up knowing how to respond to his wife, or how to be a loving father to his children? Obviously, he can't. He is more likely to feel uncared for, angry and genuinely confused by his simultaneous urges to be affectionate with the people closest to him, combined with a great fear of touch due to the messages he received in childhood that touch is dangerous and is not available to him.

What is a teacher to do? According to Barrie Beth Hansen, a music teacher in Takoma Park, Maryland, touching a student for *any* reason means putting one's job on the line as a teacher, and is not worth the risk. She explains:

> *The days of the spontaneous hug, the firm disciplinary hand—the days when the school was an extension of the home—are long gone...We live in a litigious society, with schools and educators frequently a prime target of lawsuits. Almost daily, charges of abuse against adults who work with children are splattered all over the media...Public accusations of physical abuse of children on the part of adult caregivers to children have reached almost witch-hunt proportions—a level totally unknown in even the recent past.*[7]

Hansen's rationale, that educators must protect their futures, is a valid one, and she encourages them to develop new ways of expressing their concern for students. But will a gold star, a piece of candy or the words, "it will all be all right" ever replace the comfort of a teacher's arms around a young child's shoulder when he or she is upset, or holding the child's hand when he or she is quivering from a sprained ankle or a scraped knee?

Mike McKay, a fifth grade teacher from Illinois, agrees that physical contact with students is risky, and that the teacher is subject to criticism, or worse, for engaging it, but maintains that for some students, their teachers' affection may be all that the contact they ever receive. "As teachers," he explains, "we're confronted daily with students who need caring, understanding and expressions of support. We all know students who desperately require nurturing." He further emphasized that the key to touch with students is discretion—knowing oneself, one's students, their parents, and being accountable for one's own actions. He reflects:

> *If a hug, a pat on the back, or even a "high five" had even a little to do with that feeling (that I have made a difference in their lives), then I used good judgment, discretion, and my heart. I've learned to use my heart more. It's making me a better person and, therefore, a better teacher.*[8]

Laws regarding touch exist not only for teachers, but new laws are being implemented for students as well. Says Kathy Seligman of the *San Francisco Examiner* Staff, "Labeled 'teasing' or 'mischief' since boys and girls first started going to school together, acts like these are being reclassified for a whole generation of California students. Now they have a legal name—sexual harassment."[9] Unfortunately, warned Nan Satein, a leading researcher on sexual harassment who works at the Center for Research on Women at Wellesley College, the new laws don't require the schools to teach about sexual harassment, although students from the fourth grade up can be held legally responsible for this.[10] What are the psychological implications for children when six-year-olds such as Jonathan Prevette are accused of sexually harassing their school classmates? It is worthwhile to consider what types of messages are being conveyed to children about the sacredness of their bodies when they are being taught about sexual harassment even before they can read, and before they become comfortable with the most basic boundaries concerning touch with their friends. It is also useful to look at what kind of role models children are learning from, that would result in fifty-four percent of young women at a California high school reporting that they had been harassed?[11]

Similar laws exist for psychotherapists and counselors. Imagine if, as a therapist, you decide to give your client a hug after a particularly moving therapy session, only to receive a court summons many months after the therapy has ended for having touched your client inappropriately.

In my training to become a Marriage, Family and Child Counselor in California, we were required to learn about the laws that prevent all state-approved psychiatrists, psychologists, counselors and trainees from touching their clients—we were even warned about the physical contact that inevitably occurs when payment is exchanged! I left that class with a terror of touching my clients, backed by the "no touch" rule at the counseling center where I worked, only to be subsequently faced with what it felt like not to be able to touch clients whom I deeply cared for and who were in a great deal of pain.

I have come to know of more and more qualified therapists who have opted out of state certification because of the limitations it imposes on them. I know other certified therapists who incorporate wise and discriminating touch into their practice with full awareness that they are breaking this law, yet who have decided that the healing that can ensue when touch is used appropriately is the higher principle. The flip side of this issue, of course, is that the client is in a very vulnerable position in the therapist-client relationship and there are therapists (and many of them) who abuse this power to satisfy their own selfish desires in the name of "healing."

Medical doctors, nurses, social workers and daycare professionals alike are subject to an increasing number of these laws. Those whose job it is to help people in need are facing the greatest restrictions on their freedom to give it, and the most severe consequences when they choose to do so in spite of it. According to one counselor who facilitates a children's grief group at her local hospice, "I know there is a danger of lawsuits for touching these children, but I'm just not going to sit across the room from a six-year-old who is crying about her mother dying of cancer and say, 'I understand how you're feeling.' I made a decision: I touch the children and take the risk."

Others decide not to, and not because they are uncaring. They are often people who have their own families at home and who are unwilling to risk a lawsuit for touching a patient or client inappropriately, in spite of their best intentions. Professionals who are dedicated to their jobs are struggling with issues of how to do their jobs in a way that will make a difference to their clients without breaking the law.

To understand the complexity of the situation more clearly, it is worthwhile to consider why these new laws, such as the ones forbidding teachers to touch their students under any circumstances, are made. They are not made only to protect the child, the hospital patient, or the psychotherapy client, but are primarily created to protect the state governments and the large insurance companies who will have to pay the cost if their customers are accused of touching their clients, patients or students inappropriately. It is *business.* These laws are made because, having been raised in a culture whose moral integrity is full of holes, people have been willing to take advantage of the abuse scare for personal and economic benefits, irrespective of the cost to the therapist, doctor or baby-sitter. Again, I am not saying that *most* reports of abuse are not accurate. An eight-year-old girl has little to gain by telling the school counselor that her male gym teacher came into the locker room and was watching the young girls undress. A fourteen-year-old boy is hardly going to benefit from the rejection and vindictiveness he is likely to face when he tells a social worker that his father has been molesting him on their weekend camping trips since he was nine years old. Then again, a secretary who has repressed memories of childhood sexual abuse *is* going to gain something when instead of working through her memories she files a $20,000 harassment suit against her boss for pinching her arm.

We have gotten ourselves into a difficult bind, and there are no easy answers. There is undoubtedly a need for laws and regulations that will help protect the victims of abuse, yet at the same time, these regulations must allow people the freedom to express simple gestures of care and affection. On the one hand, we must take care of ourselves in pragmatic ways, thus refraining from touching others in situations that clearly present a risk to us, and on the other hand, we must take care of ourselves and others by initiating physical contact, or by bringing intimacy and realness into relationships when that is what is needed in terms of help and support.

In an ideal situation—having been raised with lots of affection and touch, and having our requests to not be touched respected as well—we would be able to clearly and immediately assess a situation in terms of knowing when and how touch could best be used. We wouldn't have to think about it as much as we would intuitively know both who we were dealing with, as well as our own motivations behind the touch. Since we were not raised in an intimacy-free environment, we would be able to bring a tremendous amount of caring and reassurance to others simply by our presence, our quality of attention on the other, or a simple glance. We would be unafraid to touch others when it was needed, trusting our sense of the situation and, most importantly, trusting ourselves.

However, few people know themselves this well. As will be discussed subsequently in this chapter, particularly if we have been abused ourselves, but even if the abuse we endured was only that of being raised in a culture of abuse, we are largely out of touch with ourselves. It is not just "society" that is paranoid and making laws that prevent intimacy. That *is* happening, but it is we, too, who often don't know the difference between a friendly touch and a seductive one. We often don't know if our request for our child to sit on our lap is motivated by a genuine desire to spend some intimate moments with him or her, or if we ourselves are feeling lonely and are looking to him or her for our own comfort and reassurance.

The answer lies only in our ability to use our capacity for discrimination in deciding when to touch and when not to touch, in our willingness to experiment—taking small risks, like offering a supportive touch on the arm to a crying child or client, while paying close attention to how our gestures are responded to. Meanwhile, if children are not going to receive affection from their teachers at school, if clients are not going to receive contact from their therapists, and if medical patients may never be touched by anything other than a sterile gown and a latex-covered hand when they go into the hospital for surgery, the responsibility to give affection and caring rests entirely and unequivocally on our own shoulders within our homes and among our families and friends. Insecure or not, vulnerable or not, fearful of risking or not, it is the ordinary man or woman—you and me—who, if we wish to be a part of the solution instead of part of the problem, are going to have to do the touching, supplying the genuine affection so necessary in an increasingly impersonal world.

THE ROOTS OF VIOLENCE

If it concerned only
you
one might say
"What of it?"
But this sadness, this despair
which comes from your mother
who got it from her mother
who inherited it from your great-grandmother,
this mountain of sorrow which has been building up
from one generation to the next
since the beginning of time...[12] - Frederick Leboyer

The abuse of children is due to feelings of unlove in the abuser, stemming directly back to the wounds of their infancy and childhood.The abusive adult is the man who as a child was left abandoned in his crib at night where he lay screaming for his mother who never came until morning. It is the woman who, as a young girl, was taught that her value lay in her beauty and sexuality; as an adult she will go from man to man, seducing them until they fall for her and then dropping them on the spot. It is the woman who as an infant was smacked by her mother when she reached for the breast to snuggle, because her mother found nursing disagreeable; she has become the embittered third-grade teacher who expects her students to have the intellectual capacity of thirteen-year-olds, and shames them when they do poorly on their examinations.

Abuse is Passed Down from Generation to Generation

"What I do to you is nothing compared to what my father (or mother) did to me," are the words often uttered either aloud or in the silent unconscious of the abuser as he lifts the belt or the hand that will turn him into his father, or as she turns up the television to drown out the sounds of her baby's cry in the next room, just as her mother did. They express the incomprehensible helplessness and devastation of childhood wounds that are passed down from generation to generation in the form of aggression.[13]

Many people know that abuse gets passed down from parent to child, but most people do not grasp that the family heirloom of abuse can often be traced back many generations.[14] Most mothers who have allowed their husbands to abuse their children while they stood by, or who have busied themselves in the next room while the abuse was occurring, might be surprised to find that their mothers and grandmothers had done the very same thing.

Most people fail to understand the sheer power of behavioral conditioning—that the precise traits that they inherited from their own parents will be directly responsible for the most infinitesimal aspects of their child's personality. Genes are the least of what parents pass on to their children. Thus, when there is abuse present in any given household, the children present, particularly the young ones, will not only feel the effects of the abusive act itself, but will breathe in and absorb through their skin the whole toxic environment in which it occurs.

> *Most of the people in prisons—and many of those in mental hospitals—are there because of the unspeakable, unloving conditions of their early lives.*[15]
>
> - Roderic Gorney

The origins of abuse can be dated back to the species' primary instinct for survival. When it comes down to it, most people would kill rather than be killed. Unless something deeper and more unconscious was prompting them, young men of clear minds could not willfully volunteer to go to war and murder other people in a Third World country that could never possibly harm their own country, nor would they beat up their wives for not having a hot dinner ready when they come home late from work after a visit to the local bar. Mothers would not try to convince themselves, just to save their marriage, that their husbands weren't abusing their children *"that badly."* Individuals act like this because a primal instinct for survival is triggered—the soldier thinks that Daddy-government is finally going to be proud of him if he stands up and fights like a man, and will shame him for being a wimp if he doesn't (and he desperately *needs* his father's approval). The woman who allows abuse to continue in her home is terrified that if she refuses to "be a good girl and keep your mouth shut," that Daddy-husband will abandon her and that she and her kids would be left without food and a place to live. The adult is not thinking and feeling these things consciously. The unexamined life is run by these primal fears. The adult is acting this way because he or she unconsciously believes that if they do not, they will literally die.

It becomes increasingly obvious that, as adults, not only our failure to share touch and affection, but our tendencies toward violence, excessive control, depression, etc., are sustained by our unwillingness to fully feel the hemorrhaging wounds of our childhood. Innocent, essentially pure and good individuals are so fearful of their pain that they must scream at their children or train them to be passive and well-mannered automatons so that they do not have to see in their children the innocence and joy that they as parents have lost.

Abuse is Not Genetic

There is no chromosome for abuse. Conditioned? Certainly. Learned? Of course. But not genetic. As will be discussed below, abuse has been documented to have a strong chemical component at its base, but I assert that this chemical component is no stronger than the force of conscience. There are a fraction of cases in which the abuser does not know what he or she is doing (otherwise known as amnesia), but in all other cases, the abuser knows *exactly* what they are doing while they are doing it. It is true that strong emotions overcome people, and they do things they will later be shocked to reflect upon, but in the times in between, when sanity is the overriding force, there are many avenues by which to pursue help in order to prevent the recurrence of the abuse. *Conscience is the stronger force.* When conscience takes the reigns of the will in its hands, it can overrun all genetic, emotional and chemical tendencies toward abuse. I believe that people who were abused, but who don't abuse their own children, are no longer victims of that abuse. Abuse leaves the psychological tendency to perpetuate that abuse in its wake. Those who have been scarred by abuse, but who are able in their own lives not to perpetuate that abuse—even subtly—have overcome their childhood wounds enough to make a break in the generational pattern.

A client, who had confronted his parents following an intense process of recalling and reliving early memories of physical abuse, came into my office very agitated one Monday morning after having spent the weekend with his parents:

> *I couldn't believe it. My father sat me down in the living room to have a "serious talk" with him. He*

went on to say that he had researched the history of mental illness in our family and wanted to tell me about my great Uncle Les who was schizophrenic, my maternal grandmother who was a "little bit off her rocker," and my third cousin, Joe, who was mentally retarded...I gasped when I saw where he was going with this. He wanted to convince me that I was crazy for thinking I had been abused. "You know," he continued, "these kinds of things are genetic, so I thought it would help you understand your need to make up these kinds of stories." Calmly and without saying a word, I stood up and I walked out of the house.

Abuse May Have a Chemical Component

Recall from Chapter Two that a failure to bond in childhood inhibits the natural occurrence of certain chemical functions that are necessary for the normal developmental maturation of the brain. James Prescott says that the neuronal systems of the brain are responsible for mediating pleasure, and that in the unbonded individual, the development of these has been stunted, thus failing to regulate the chemical aspects of depression and rage, and to balance them through pleasure. "The failure to integrate pleasure into the higher brain centers associated with 'consciousness' (frontal lobes) is the principal neuropsychological condition for the expression of violence, particularly sexual violence."[16] It becomes a plausible hypothesis that among the chemical processes catalyzed by bonding is also that which is needed to form a mature conscience in the individual.

Perhaps the most interesting of Prescott's findings emerged from his cultural studies on violence. He documented statistical evidence to back the fact that the presence or absence of touch is not only the distinguishing factor between a healthy and well-adjusted individual or a violent one, but also how cultural practices regarding touch are responsible for creating either violent or non-violent societies. He developed the means to predict, with 100 percent accuracy, whether a culture would be violent or non-violent simply by looking at whether the infant was carried on the body of its mother or caretaker (this factor alone resulting in 80 percent accuracy), combined with whether or not adolescent sexual activity was permitted or punished.[17] He explained, "Those societies which give their infants the greatest amounts of physical affection were characterized by low theft, low infant physical pain...and negligible or absent killing, mutilating, or torture of the enemy...The warmth of human touch and security of body contact are without question the most effective way to reduce violence in our culture."[18]

Thus, there is a direct relationship between touch and violence. When an individual has been adequately nurtured in childhood, he or she is highly unlikely to act out aggression in his or her environment.

Most Children Would Rather Have Abusive Contact Than No Contact

Jonathan Kozol explained:

> *Children long for this—a voice, a way of being heard—but many sense that there is no one in the world to hear their words, so they are drawn to ways of malice. If they cannot sing, they scream. They are vessels of the spirit but the spirit sometimes is entombed; it can't get out, and so they smash it!*[19]

Ironically, most children would rather be abused than be neglected or abandoned. What they really want is to be loved, held and given affection and for their parents to take interest in them. They want their sexuality to be supported and respected. They want evidence of their parents' love in the form of their attention. But when it comes down to it, abusive contact is often preferred to isolation.

As mentioned throughout the book, tactile stimulation is necessary for the growing body to develop and for the developed body to stay in balance. A child who is raised in a touch-rich environment will naturally seek contact, but even the child who is abused or neglected will seek this contact. The child's body will naturally look for what it needs to stay chemically balanced. The child may not know why she jumped into her mother's lap, or suddenly found himself in a screaming battle with his father, but unbeknownst to the child, the body had signaled to the brain what it needed in order to maintain balance. Many children are accused by their parents of provoking abuse; perhaps this is so, but the child's underlying motive is a need for contact rather than a desire to be abused.

When there is abuse, there is hope, in an odd sort of way. When the parents are unaffectionate with the child or absorbed in adult-conversation or activities, the child is stuck in isolation, whereas when there is abuse, there is action. The child has a chance to make a stand for himself, to try to be heard, to make a point—even if it entails further abuse. When nobody else is noticing him, he at least stands to gain something in terms of self-esteem for his willingness to advocate himself.

An example of this appears in homes where the father abuses the mother, and she receives it passively. In these instances, it is not uncommon for still-innocent young children to provoke fights with the father in an effort to "save" their mother by demonstrating to her how to fight back, or by diverting the abuse from her, or in an attempt to show her what it means to put another person (i.e., the child) first.

French physician and psychologist Dr. Alfred Tomatis further suggests that stimulation is sought through sound. There is a strong correlation between touch and sound. The vibratory effects of sound travel through the ear and stimulate the same centers of the brain as does tactile sensation, thus accounting for why infants, even in utero, immediately relax to certain types of music (e.g., classical, lullabies), and show disturbance in response to harsher forms of music. (For more information, see "In the Womb," Chapter Three.) The unbonded teenager still craves this same touch. Thus, according to Tomatis, he or she will seek out loud and highly stimulating music (e.g., the now-popular heavy metal and rave music) in order to get the stimulation that his now-undeveloped system is craving. His ideas offer a plausible theory as to why today's youth culture has turned to such harsh forms of stimulation in search of "entertainment."

On this point, psychologist Erich Fromm suggested that if children can't get their strokes from life-giving situations, they'll get them from life-taking situations. If people, and not only children but all people, are not being touched by life, they need to be touched regardless, so they'll cut themselves with razor blades or touch themselves with body piercings. Adults will touch themselves with gluttony, or tobacco, or drugs, or by hitting their spouses. They will touch themselves by hurting the people they love, as this is what gives them the strongest pain—and pain is at least a feeling in an otherwise numb body.

The abusive adult is merely an abused child grown old, unless per chance he has done his own healing work. If as a child he learned to get contact through pain—either by provoking it or being the recipient of the other's provocation—he will automatically enact this as an adult. A client told me the story of how often, at the dinner table, his father would start screaming at his mother or at the children when he felt that his point of view wasn't being respected, or when his children weren't interested in him trying to teach them about his law practice. This is no different than the young child who, unable to get Mommy's attention or affection, throws a temper tantrum.

This principle—the need to get attention, even if abusive attention—applies to sexual abuse as well, whether it be father-daughter, uncle-niece, or teacher-student abuse. The perpetrator (often unconsciously) uses to his advantage the knowledge that the child is raised in a generally abusive and unhealthy environment (often of his own creation), and therefore craves affection, acknowledgment and attention. Therefore, if he plans his attack well, he can possibly get full consent from his victim. For example, a father tells his son, "You're such a beautiful young boy. More beautiful than any woman, and what I need from you I can't get from your Mommy." Or a teacher tells his fifth-grade student, whom he has "detained" after class, "You are the most attractive girl in the whole class. I want to show you just how special you are." Remember, a touch-starved, unbonded child is at the effect of an inner desperation. Although deep in the body they know that something is very wrong, they are finally hearing the words they have longed to hear, and then being touched, held, and made to feel pleasure. They feel special. They feel important. They feel touched. This does not happen in all cases, but is a poignant example of how someone in an already touch-starved state can become easily manipulated into further abuse.

Evil Exists, But So Does Conscience

In his book *People of the Lie*, best-selling author M. Scott Peck risks the radical proposal that evil lies at the base of many crimes of abuse. Whereas we normally associate the concept of "evil" with gross images of violent actions, Peck suggests that evil is a force dictated by a lack of conscience that can manifest in "ordinary" parent-child relationships, and in subtle but definitive ways.

As a child of nine, I recall seeing a news report on the trial of a mass murderer who had escaped from prison, and had subsequently murdered several more victims, many of them children. The news showed him waiting to receive his sentence, sitting in a courtroom packed full of people. When the judge asked him publicly if he had any regrets about the crimes he had committed, the murderer looked the judge straight in the eye and said, "None."

I had seen evil. Although this murderer was speaking his conscious reality, both then and now I do not believe that conscience did not exist somewhere inside of him, no matter how deeply hidden.

Supporting the aspect of Peck's theory that evil does exist, I would take this back one step and purport that *one is not born evil,* but that *evil is created.* It is highly unlikely that this murderer was born to a conscious mother and father, was nurtured in the womb, was welcomed into the world in a loving way, was breast-fed on demand, carried around in his first two years, allowed to share his parent's bed, and given lots of affection as a child. If Adolph Hitler had been raised in the environment just described, I believe that he *would not* have murdered six million

Jews—he *could not* have, because the conscience that exists in every individual, if only as a minute speck of awareness, would have been awakened in him, and would not have allowed him to commit these crimes. Touch, affection and honest care and consideration must be communicated from conception onward. The force of evil—capable of murder, incest, and even global destruction, as well as being responsible for the insidious ways we invisibly denigrate even those we love most—is the result of nothing more, or less, than the human failure to love properly.

In the initial version of James Prescott's 1996 letter to the editor of the *New York Times* (which was significantly edited before printing), he set forth the following challenge:

> *The legal-judicial and criminal justice system is challenged to find ONE murderer, rapist or drug addict who has been breast-fed for "two years and beyond"—the recommendation of the World Health Organization—in any prison, jail or correctional facility in the United States.*[20]

Abusive Tendencies Exist Within Each of Us

If you have ever been willing to be ruthlessly honest with yourself in the moment in which you have gained power over another individual (in a fist fight, an argument, or even a so-called friendly debate), you may have been surprised to find that you found a certain satisfaction, or even pleasure in it. The feeling is no different from what is felt in a football game, a tennis match or a game of bridge at the moment when you know your opponent doesn't have a chance, or from the thrill of thousands of spectators at a bullfight when the matador pierces the bull, or the look on the lion's face when he has caught a gazelle. There is satisfaction in the kill. We enjoy the power of dominance, the taste of blood. In it we feel strong and invincible. Our survival is vindicated—we have the illusory experience of momentarily conquering our death. Many people will do anything for this.

The abuser is not other. The capacity to abuse exists within each of us. If you are using this book as a tool to explore yourself, you will know as well as I do that nobody in contemporary Western culture—save perhaps the rarest of rare—escapes the reality of this collective human wound. Our capacity to abuse is the curse and the blessing of the human condition. This wound is the curse because of the suffering it creates, and the blessing because in the understanding of it lies the seed of True Compassion. The sooner we are able to recognize that abuse is inside of us, the sooner we will be able to see it for what it is, instead of running from it, fearing that it is something terrible about *us,* or that *we* are bad. When we stop judging ourselves for our potential abusiveness, we begin to understand how it works in us, and we begin to understand how it works in others. We can then cease to judge others, and cease to defend ourselves from the intimacy we so crave. Thus abuse loses its grip on us.

Abusive tendencies exist within each of us, but everybody is not an abuser. We cannot help but to unknowingly hurt people in small ways as we move about our daily lives, but abuse is a function of action, not feeling. The more unaware a person is of their abusive tendencies, the more likely they are to act them out. Becoming aware of these feelings does not mean acting them out. If you have to, you can paint them, write them, ride them out on your bike, or punch them into a punching bag, but generally you let the feeling be, allow the intensity of it to teach you not only about yourself, but about the greater human condition. When we understand that the problem is not "other," we simultaneously understand that the solution is not "other."

HOW WE COPE

Defense Mechanisms

The field of psychology uses the term "defense mechanisms" to describe the specific forms of thought and action that abused individuals adopt in order to keep the pain of trauma out of conscious awareness and to protect themselves from experiencing the full impact of the abuse. Understanding how defense mechanisms work allows us to come into a fuller appreciation of the consequences of the wrong kind of touch, and what an individual attempting to heal this abuse is faced with.

Dissociation

Commonly, when a child is being sexually abused, her consciousness will take leave of the body, traveling up onto the ceiling or across the room. She literally separates from her body, becomes identified with her mental consciousness, and thus affirms, "*I* am up here. The body that is getting abused is down there."

In order to sustain this belief in adulthood, such a person must remain disconnected from her body, for her body "knows" the abuse she suffered. As an adult, she is likely to not really be *at home* in her body. Of course she walks and talks and eats just like everybody else, but she will often appear to be spaced out, unfocused, and have an awkwardness in her movements. When I am with a client and begin to feel disoriented or spaced out, I look to see if she is present "behind her eyes," that is, to see if somebody is there. Nine out of ten times, I find nobody home. "The hitting, the screaming, and smothering affection made me want to go somewhere else," said one young woman, "so I created a bubble in which to live in my head." Unfortunately, greater than the pain of the abuse itself is the individual's lifelong loss of connection with her body, for in keeping herself detached from the pain in her body, she also removes herself from the pleasure and nurturing qualities of touch.

Numbing

> *In order to cut off unpleasant feelings and sensations, the child will hold his breath, suck in his belly, and immobilize his diaphragm. He will lie very still to avoid being afraid. In short, he will "deaden" his body in order not to feel pain, and by these means abandon reality.*[21]
>
> \- Ashley Montagu

A boy may unconsciously numb the surface of his skin as a means of self protection because his father regularly whips him with a belt. That same child may be seen outside in winter in shorts with no shirt, commenting that he doesn't feel the cold. A girl who has been sexually abused may lose all sensation from the waist down. She is able to feel external stimulation on the skin, but cannot feel the life inside of her body. One of my massage clients said she always felt like she was "walking around on somebody else's legs"—that she knew her legs were attached and would carry her around, but that they didn't feel like a part of "her."

A more extreme example of body numbing is illustrated in a story told by a friend of mind who is a nurse. An obese thirty-two-year-old woman, complaining of intense stomach cramps, admitted herself to the hospital where my friend works. Within hours of entering the hospital, she gave birth to a baby boy. Utterly shocked, this woman had carried a baby in her body for nine months and had never felt it!

Fantasy

Fantasy is another way that victims of abuse cope. The environment they live in is too threatening, so they absorb themselves in their own make-believe world, or in fictional worlds created by others. Whereas in young children, fantasy is a normal and healthy aspect of development, adults who spend the majority of their time in front of a computer or a television set instead of with family and friends, or absorbed in trashy romance novels, are using these activities as a crutch.

Disavowal

Many children who have been abused *disavow* their desire for touch to protect themselves from being hurt again in the same way. They may lose their desire for contact altogether. Somewhere early on they make the vow, "If this is what it means to be touched, I don't want it again—*ever!*" Or, "Love hurts too much, I'm *never* going to let somebody come close to me again." Such vows made by children are more powerful than most people would imagine, and often have lasting effects. They become part of the individual's unconscious, but are lurking nearby and readily available whenever any possibility for contact arises. For example, the woman who claims, "Touch with one's husband is fine, but among girlfriends is really inappropriate," probably grew up in a family in which her father gave her ample and thoughtful affection, but where her mother did not cuddle with her or express her affection freely. For the man who says, "I just don't like to be touched, it's as simple as that," it is probably not so simple. He was either neglected, emotionally smothered, or physically or sexually abused.

Fortunately, the soul, or Self, can also make vows. One man I know recalls the precise time at which his True Self went away because of the pain of the abuse he endured as a young boy. However, at the same time, he made the vow to himself, "I swear that when I grow up and finally get out of here that I'm going to do everything I can to get my real Self back." And he did. He is now a thriving thirty-five-year-old man with an excellent career and a wonderful family of his own.

The Multiple "I's"

In my own experience, I have found that aspects of the spiritual philosophy of Russian teacher, G.I. Gurdjeiff, have very practical psychological use in understanding and healing the consequences of abuse, and therefore opening oneself to a life full of touch and affection. In particular, his notion of the "multiple I's," though originally intended as a description of the spiritual development of the human being, can be applied to the psychology of the abuse victim. He says:

> *It is the greatest mistake to think that man is always one and the same. A man is never the same for long. He is continually changing. He seldom remains the same even for half an hour. We think that if a man is called Ivan he is always Ivan. Nothing of the kind. Now he is Ivan, in another minute he is Peter, and a*

minute later he is Nicholas... You will be astonished when you realize what a multitude of these Ivans and Nicholases live in one man.[22]

Many individuals, particularly those who have been traumatized, are not aware of the complexity of their personalities. According to Gurdjeiff, multiple "I's" exist within every individual—a series of distinct identities which together constitute that which the individual thinks of as "himself" or "herself."

An example of this is the man who makes a decision to start getting up early every morning to exercise. However, when his alarm goes off in the morning, he presses the snooze button and goes back to sleep. Frustrated, he restates his intention the following day, only to remain in bed once again. Why is this so? The "I" who made the decision to get up, and the "I" who woke up in the morning, were not one and the same. The "I" who decided to get up and exercise really would do that, and the "I" who wouldn't get up, really wouldn't! Factors such as clouds or sunshine, heat and cold, as well as emotionally threatening situations, can call up entirely different sets of "I's."

Though the reality of the multiple "I's" is not commonly recognized for what it is, it is a recurrent theme in the work of artists including painters, musicians and sculptors. The lyrics to the popular Eurythmics song, *This City Never Sleeps*, captures this multiple "I" phenomenon:

I can hear the sound of the underground train
Though it feels distant thunder...
You know there's so many people living in this house
And I don't even know their names.
You know there's so many people living in this house
And I don't even know their names...

Oh some days I can almost hear them breathe
If I listen then I hear my own heart beating...
Guess it's just a feeling...[23]

A similar dynamic occurs in relationship to touch. Let's use the example of a fairly new couple going out on a date. Perhaps the man picks the woman up and gives her a quick kiss, and the woman is slightly annoyed. Her "I" that answered the door was not immediately open to touch. They go out to a nice dinner, have some wine, and walk out of the restaurant arm and arm—they are each operating under the auspices of an "I" that feels safe, comfortable and relaxed. Afterwards, at the movie, he reaches out to hold her hand, but the "I" whose hand he takes is already different, feeling self-conscious about the intimacy that was shared after dinner and afraid of what will happen if she gives into it, so she quickly excuses herself to go to the bathroom, and when she sits down again has her hands covered under the sleeves of a sweater she has put on. Yet, as the movie, a romantic drama, unfolds, an affectionate "I" appears and shyly she reaches for his arm. It is he who is taken aback now. In response to her previous clues, a defensive "I'm not interested / I don't need you", "I" has taken over in him. She retracts in reaction to his response, as her "if you don't want me I certainly don't want you 'I'" comes to her rescue. The movie ends with one of the lovers dying, making a tiny crack in the woman's protective "I", and a softer and more vulnerable "I" emerges in her. On the way to the cafe, she takes a risk and tells him how much she enjoyed their time together and that she hopes it continues into the future. His "I" immediately shifts into a softness complimentary to hers and they enter into a romantic mood together.

Among individuals who have been abused, the phenomenon of the multiple "I's" is particularly pronounced. In order to cope with the trauma, the mind of the abuse victim often fragments into several "I's," or sub-personalities. Many of these "I's" do not even know about the abuse. The far end of this spectrum is known as "multiple personality disorder," in which the individual has a series of "I's" that are not only totally distinct from one another, but totally unaware of the existence of the other "I's." This individual may say or do things as one "I" that he will not even remember when he is another "I."

Multiple personality disorder is the exception rather than the norm, however. When people do not understand that multiple "I's" exist within everyone, they are likely to believe that there is something very wrong with them when they do not understand the range of moods and feelings they experience within a given day. They are also unaware of why in some moments they are very open to touch, and to giving and receiving affection, and why in other moments they are rigid and adverse to touch. Such individuals often believe themselves to be "crazy," when instead they are experiencing a phenomenon common to all human beings to a greater or lesser degree.

Touch is the Integrating Factor

Touch is a common meeting ground for the many "I's." When a trusted individual is present, and the body understands that there is no imminent danger, the defenses can relax, and the dissociated, disintegrated and split off "I's" can begin to come together. What people call a feeling of openness is often simply a relaxation of the defenses that are ordinarily keeping the separative "I's" intact. Safe, intentional and conscious touch by caring individuals can provide this healing.

The touch that heals may not be physical touch initially. Remember that the abused or touch-starved person may not be ready to receive physical touch right away, even when she knows that is what she needs. Her early wish for contact was brutalized, shamed, and often turned against her, and she is still wary and uncertain. The person may need "talk touch," listening touch, the touch of attention, of a steady gaze or of shared silence and presence. She may be able to feel touched at some moments and not at others. She may ask for touch and then freeze up when it is given to her. She may need the touch of patience. Since the skin was the scene of the crime (particularly in cases of extreme violation of the body, or the neglect of its needs), only by returning to the skin—not only facing the terror of touch, but transforming it from a source of fear to a source of pleasure—will there be a transformative effect. When the person who once disavowed touch has not only learned to tolerate it but to again desire it, she has cleansed the wound and is well on her way to a sense of well being.

SEXUAL ABUSE—THE TWISTED TOUCH

> *The absence of loving touch and the abundance of sadistic sexual touch has badly wounded me. Man does not live by breath alone.*[24] - Greg Campbell

When a person is sexually abused, his or her physical experience of pleasure often becomes directly linked to emotions of terror, betrayal and "badness," and thus the body begins to experience loving touch as anxiety-producing, and gestures of affection as warnings of betrayal. The body and brain operate by a system that can be likened to complex wiring, and for the victim of sexual abuse, it is as if somebody twisted the wires that hooked up the normal pain and pleasure responses, thus severely impairing the abused person's organic capacity to give and receive touch and pleasure. You will often find these individuals to be highly intelligent and competent in many arenas of life, while at the same time extremely self-deprecating and undeveloped in other ways.

> *I crave sex just like everybody,* began Sheri, a 32-year-old sculptor, *but when my body starts to feel good, I am overcome with a sudden exhaustion and can't even stay awake, or I have a sudden allergy attack, or I am overwhelmed with feelings of hatred and rage at my husband, even though I love him. Sometimes when we have sex, I feel like I am being raped, even though he is gentle and kind. The torment inside my mind at these times is indescribable.*

The sentiment expressed above is affirmed in the lives of countless other women and men. What is common to all of these individuals is that they coped with their childhood abuse by learning, in some way, how not to feel. Unfortunately, this defense becomes stabilized and automatic in the body, and when they reach adulthood, even if they are fortunate enough to find themselves in a completely non-abusive circumstance, the defense system will oftentimes still not be able to discern the difference between a healthy and an unhealthy environment, and will continue to operate as though it was still in an abusive situation.

These individuals have lost their capacity to receive touch, and being so removed from their bodies, lack the sensitivity needed to touch others with love and gentleness.

Instead, they grow up and repeat the cycle—men who were sexually abused will usually abuse, and women will find someone to abuse them. That is, they repeat the cycle until they decide to break the cycle. For, as will be discussed in Chapter Five, it is the touch that was once used as a weapon to wound them that can now be turned into the medicine that heals them.

Abuse leaves the individual distrustful of touching others. A letter to the editor of *Mothering* magazine from a woman who had been sexually abused as a child reads, "My son Noah is eighteen months old, and while changing his diaper or bathing him, I have often wondered: 'Is this all right?' 'Am I touching him too much?'" Similarly, a man in his early forties shared, "When I was a child, my father (who was a schoolteacher) was found guilty in court for sexually abusing several of his students as well as two of my cousins. Now when I am being sexual with a woman, I'm afraid I'm being abusive, or that I'm touching her like my father would."

Those who have been abused and yet *do* wish for contact will often be very awkward and insecure about touch, approaching it with so much anxiety and apprehension of rejection that they create a situation that will sometimes (or often) backfire on them. Not only adults who have been physically or sexually abused as children, but those whose parents did not hold them, stroke their hair, or give them hugs, will not know how to touch another person with ease, or to reciprocate another person's gestures of affection.

Many individuals find it extremely frustrating to discover within themselves many or all of the symptoms of abuse listed here, but with no memory of any kind of dramatic abuse. Their childhood may appear quite normal and mild compared to the kinds of abuses that are now so commonly spoken of. Perhaps their mother was depressed or their father was a closet drunk. Or maybe their parents were divorced and their mother struggled financially and could not afford to feed her child well. As mentioned early in this chapter, abuse is not being discussed here in terms of blame, but instead in terms of how it affects the individual's capacity to give and receive touch. Therefore, it is less important that a person know exactly *why* they have difficulties around touch, or *who* is responsible for it, than it is to acknowledge the fact of *what is true for them,* and to seek ways of returning to feeling.

We continually underestimate the sensitivity of the child and how deeply he or she is influenced by their environment. In many cases, simply being raised by a mother who, unbeknownst to the child, had been abused by her own mother, that child, by virtue of his attunement to his mother, can easily absorb her devastation and fear of others, thus becoming skin-wary himself without ever having an inkling as to why.

The Point of No Return

There is a *point of no return* when it comes to abuse. Many people have had the experience, in the midst of their attack on another, of suddenly meeting the terror in the victim's eyes—a look that shocks the attacker into the realization that they have gone too far, that something has been broken. Such a glimpse of reality can be shattering. The individual recognizes just how much he or she has hurt another person—for deeper than the damage to the skin, the sexual organs, the pain-pleasure circuits and the emotions, is the soul wound. The soul can be broken, and to repair a broken soul, if at all possible, often requires many years of hard work—all for a momentary lapse of consciousness or a selfish indulgence on the part of the abuser.

As a therapist, many clients have said to me, "My mother told me that I was a 'mistake' and that she wished I had never been born, but I know she didn't *really* mean it." Or, "My father got carried away and locked me in the closet—but it only happened once." Once can be more than enough, however, to bring someone to the point of no return. The soul is not broken every time (some escape irrevocable damage), but the fact remains that when an individual slips into an abusive moment, he or she does so with no guarantee or insurance that, "Everything will be O.K."

One client of mine, a young woman who had been severely abused by her father, knew just when her soul had been wounded. "When I was three—that's when I went away."

When a child, teenager or adult shows no need for contact—not just a hug, but a complete lack of desire to engage in real intimacy, you can be sure that he or she has lost his or her soul. When a person is so superficial that you feel like you're talking to a wind-up talking doll, and that every word you say and every gesture you make just bounces off of him or her like a super ball, it is likely that the soul beneath the facade has been broken. Although children are tremendously resilient, there comes a point for many when they can no longer tolerate the abuse and they shut down entirely—resulting in a suicide of the soul.

If you look closely, you can almost always tell when somebody has lost their soul. Soul loss manifests in the child who looks far too old for his years—not just mature, but tired, worn out and very serious. Or in the child who appears to have totally "checked out," as though he saw too much and could not bear to see any more. The adult who has been broken appears to have a hardened shell, or a suit of armor, surrounding him. You speak to them and they push a button on an internal cassette-player in response. "How are you doing?" you ask. "Fine," he responds. "Where are you going?" you continue. "Out," he tells you. And so forth. The tape has variations, but the root cause is all the same. You feel a heaviness and deep depression when you meet the person who has sold out.

> Michael is such a person. Raised by a rageful father who beat him and emotionally degraded him (a father who himself was raised by a man who brutally beat him and molested his younger sister), Michael's soul was broken not by being sexually abused himself, but through the years of listening as his father molested his baby sister in the next room. Michael had prided himself on being the elder brother—his baby sister Sarah was born when he was eight-years-old. He adored her and felt very protective of her, especially because he knew what kind of home they were living in. Michael had hoped that his sister, being a girl, might escape their father's wrath and their mother's passivity. But, by the time Sarah was two, their father began to scream at her and degrade her. Then Michael would step in "like a man" to save her, claiming she was only a baby and often succeeding in gaining the support of his mother to help comfort his crying sister. However, when Sarah was three, and their father began to visit her bedroom late at night, Michael did not know what to make of it. One night he walked into Sarah's room, thinking his father might just be checking in on her, only to find his father touching her in places the young boy clearly knew were inappropriate. His father grabbed Michael by the neck, and told him if he ever

said a word about this to anyone, that he would kill him, and insisted that Michael agree. Michael consented, and in that moment his soul was broken.

Thus began many years of continued heartbreak on top of an already broken soul. Michael had made a decision to save his life, yet it had cost him the very life he was trying to save. As the abuse continued throughout the years, Michael distanced himself from Sarah, often ignoring her, and pretending as though nothing had happened. He could not bear to live with himself knowing he was not protecting the sister he loved so much, and so he cut off his love for her too.

Michael started having frequent nightmares, and would often only sleep a few hours each night. His mother received telephone calls from Michael's school teachers, reporting that he frequently feel asleep in class. Instead of receiving his symptoms as a message to her, Michael's mother sent him to therapy at age ten to deal with "his" problems. However, when even after many months Michael's nightmares did not subside, the therapist suggested to Michael's parents that perhaps it was something to do with their own relationship or a family dynamic that was at the root of Michael's problems.[25] Infuriated, his parents immediately took him out of therapy.

As a teenager, Michael numbed himself, using every spare moment to dissolve into the television set, and to take all the drugs he could get his hands on. As an adult, Michael has never had an intimate relationship. When asked how he's doing, he routinely responds, "Life's boring. That's how I like it. That way there's no surprises." This is the consequence of the wrong kind of touch.

CULTURE OF ABUSE—CULTURE OF DENIAL

> ***Cultures and societies—like parents—are not all equal in their ability to provide and receive love.***[26]
> \- Roderic Gorney

A "culture of abuse" suggests not only a culture in which abuse is rampant, but a culture in which the same elements and symptoms that characterize the abusive family system are manifest in the society as a whole. In the same way as the abused child becomes a part of the abusive family system that has molded him, so the individual members of a society become a part of, and cannot be separated from, the larger aggressive economic and social systems that are responsible for the culture in which they live. This "abusive system" will permeate all aspects of society from the very structure of the government that runs it, to the individual roles in the cultural system, to the values and ideals maintained by the people.[27] A culture of denial goes hand-in-hand with a culture of abuse, as abuse depends upon denial in order to sustain itself.

The System

In the same way that abused boys often grow up to be aggressive, angry, hurtful, disconnected and emotionally handicapped adults, a patriarchal culture (including patriarchal women!) in which abuse is pervasive will manifest these tendencies on the collective level. For example, in a culture of abuse there will be a "dog-eat-dog" ethic and a get-rich-quick obsession. (The abused boy or girl thinks: "When I make it to the top and become rich and powerful, my parents will see that I am smart and successful.") There will be a strong basis of military power and dominance. (The abused child decides: "I'm going to become so strong that nobody can ever touch me again—I'll show them who to be afraid of.") There will be high rates of crime. (The hurt child figures: "Nobody respected my boundaries, why should I respect theirs?") Rape and sexual crimes will run rampant. (The son who was either sexually abused by his father, or watched his father sexually abuse his mother or sisters thinks: "They fucked me up, now it's *my* turn.") Drug and alcohol abuse rates will be high. (The wounded child looks for any possible way to cope with feelings of grief, betrayal, rage, abandonment.) And the society will think of only its own benefit and the fulfillment of its present desires, even if that entails sacrificing the very earth that is to provide for its grandchildren. (The abused child turns his attention to himself, often becoming selfish as nobody else seems to be taking care of him.)

Meanwhile, the abused girl grows up to be passive, afraid, shamed-based, and has a low self-esteem. Thus, the feminine aspect of society allows this violent culture to sustain itself as the dominant force (as the once-abused girl marries an abusive man just like her father, though she swore she never would). In the same way that the oppressed woman feels powerless to do anything as her husband screams at her children, she finds herself equally helpless to act when the government cuts funding for education and welfare. (The abused girl, unable to fight back against her parents, gives up her power and her voice.) Because she never learned to speak up, she grew to believe that what she had to say was unimportant. Other abused girls cope with their pain by becoming so strong, so competent, and so "equal" to men that no man can have power over them again. (Thus, we now have a culture full of women in business suits or driving bulldozers while a daycare center raises their children.) And as the child who was sexually abused learns that she can get love and attention through drawing attention to her body, we now live in a culture that supports promiscuity, pornography and the objectification of women's bodies. And just as the abused girl never learned to stand on her own two feet, thus we have a culture full of skinny Barbie dolls in high heels who have lost a sense of their true femininity. Now a wealth of feminine wisdom and knowledge lies buried in the collective unconscious of the culture—its worth entirely denied.

These attitudes of abuse pervade all facets of society. It is socially acceptable to go into other countries and murder tens of thousands of people. It is sanctioned to lawfully extract money from all the people in a given country in the form of taxes in order to support technology that is capable of destroying the human race at the push of a button. Multi-national corporations which exploit immigrant workers are glorified for their economic achievement. Journalist Robert E. Fatham reports that the Humane Society can bring charges of cruelty to animals for the same behaviors that are sanctioned when directed against children in the public school system. According to the National Coalition to Abolish Corporal Punishment in Schools, the only developed countries still permitting teachers to hit students are the United States, South Africa, parts of Canada, and Australia. Most European countries began to outlaw this type of abuse in the 1800s.

A culture of abuse is one in which its members feel generally let down, uncared for and disregarded. (On the superficial level of society, everything may appear to be just great, just as the face of the abused child may show little or nothing, while her heart aches within.) The culture does not know True Touch, yet often is well acquainted with the horrors of abusive touch. And it is not only the physical touch that this culture misses, but the experience of being surrounded by trustworthy and generous people. (In many traditional villages, the community was mutually devoted to the health and economic maintenance of all of its members. Presently in our culture many people cannot afford to go to the dentist when they have a toothache.) A culture of abuse does not support the experience of living in a world that respects the individual, and in which there is an overall sense of belonging and well being.

Again, it is not that everyone within the culture is abused. Instead, the overall symptoms of abuse have begun to permeate the culture such that general feelings of aggression and helplessness, plus a fear-based drive to consume—as opposed to a feeling of generosity, balance and consideration—have come to dominate almost all arenas of modern life.

The People

> *In civilized societies...the differences among people are largely expressions of the ways in which they have adapted to the distortions in their personalities caused by the qualities and quantities of deprivation they have experienced.*[28] - Jean Liedloff

When abuse has become the norm of society, it is undetectable by its members, as there is nowhere it is not present. People have no paradigm of wholeness and peace by which to perceive the contrast of their present state of devastation. For example, the child who was raised in a household of sheer psychic insanity knows no model of sanity, and therefore accepts his terrible condition as normal, as objective, and from birth (or before) automatically begins to adjust himself to these conditions. Similarly, children born into a culture in which the underlying mood is one of aggression, competition, fear and depression, and in which the whole economic and family system is built upon this foundation, will literally cramp themselves, closing their hearts or their higher minds, or whatever they must, in order to accommodate the norm. They quickly perceive the vast gully of separation between their own innocence and purity and the aggression that surrounds them. In order to believe that they will be taken care of, children tell themselves, "Mother and father know best, and would never do anything to hurt me." Projected onto the culture, this attitude later becomes, "It's a democracy, the President is of the people, and everybody has equal rights. If I pay my taxes and register for the draft, I will be taken care of." For the sake of its survival, the child concludes that mother, father, culture, family structure, president, bank, democracy, etc. must be right, must be good, and so he recreates himself to become a part of it. Montagu elaborates:

> *...Driven to armor himself against attachment and betrayal, he (the child) may have arrived at a state in which all contact seems repellent, where to touch or to be touched means to hurt or be hurt. This, in a sense, has become one of the greatest ailments of our time, a major social disease of modern society that we would do well to cure before it is too late. If the danger remains unheeded, then—like poisonous chemicals in our food—it may increase from generation to generation until the damage has gone beyond repair.*[29]

There are many roles that the members of an abusive culture take on. Family systems theory, a widely accepted psychological model of how families operate, can be applied to the larger family, i.e., society, as well. The theory dictates that the family—whether functional or dysfunctional—operates as a whole unit, each part dependent upon all other parts for its functioning. Like a body that depends upon the heart, lungs, kidney, brain, arteries, etc. to function, and which cannot sustain itself when missing any of these components, so the family system operates—the abuser cannot abuse without a victim, and those present who do not prevent the abuse are active in their passivity. It is a common family scenario for the "man of the house" (the abuser) to hit his children (the victims) while a mother (the passive witness) does the dishes or stands by whining softly, "Don't do that." When these children grow up, they tend to repeat the cycle by becoming abusers, victims or passive witnesses.

Similar patterns show up in the members of a culture of abuse. The abusers are the people who hold power and wield it (often unconsciously) for their own benefit. Many of the power-wielders who run the country and its systems fall into this category—often they will be presidents of large corporations, government officials in departments of war, financial officers, dishonest businessmen, brokers, soldiers, policemen, criminals, etc. They become the perpetrators in the culture of violence in lieu of becoming its victims (in the exact same way as most young boys who are abused cope with their feelings of helplessness by becoming abusers, giving them the illusion of finally having power over the other, even if "the other" happens to be their own four-year-old child). This common pattern is often expressed as, "Kill or be killed," or, "May the best man win."

The victims are those who are at the effect of the societal abuser. Their voice, power and rights have been taken from them by the forces of discrimination. Included in this category are those who are African American, gay/lesbian, Hispanic, poor, homeless, mentally ill or immigrants. These individuals are largely abandoned by society, and experience the same disregard and lack of acknowledgment that the child feels from his abuser. They feel violated and stepped on by racism and inequality—being increasingly denied their basic human needs while a small, white, wealthy majority continues to grow more prosperous. Of course in these victims' minds they can see that these attitudes are ridiculous and reflect the views of society (the cultural "parent") and not themselves, but if day in and day out this is their reality, it begins to seep in, just as the child internalizes the abusive attitudes of his parents. When a whole culture is telling you daily in a hundred different ways that you do not matter, that you are less-than others and undeserving of the privileges allotted to them—it is nearly impossible not to take this on to a greater or lesser degree.

The passive witnesses comprise the majority of the members of a culture of abuse. Abuse can only occur in a system that supports it, and thus, although the present system is terribly unjust, the impotent members of society are as much a part of it as are the leaders they elect. The passive witness is the general public. Their sentiment could be expressed as, "As long as it's not happening to me, I don't want to know about it," or, "I'm too scared that if I interfere I'm going to get beat up too, so I think I'll just stay off to the side." We are all passive witnesses to some degree, even if at times we take on other roles. We want to get on with our lives, we don't want to suffer, we feel overwhelmed by the atrocities around us, it hurts too much to care, we don't want to believe that we can make a difference and so we remain detached and uninvolved. The price of the passive witness is summarized in Martin Neimoeller's famous poem, *First They Came for the Jews*:

First they came for the Jews
and I did not speak out
because I was not a Jew.
Then they came for the communists
and I did not speak out
because I was not a communist.
Then they came for the trade unionists
and I did not speak out
because I was not a trade unionist.
Then they came for me
and there was no one left
to speak out for me.

Lastly, there are the "black sheep" of the family. They are those children who did not, who could not buy into the illusion that Mommy, Daddy, country, or president were omnipotent and would take care of them.

> *he* seemed *protected...but inside: who could ward off, who could divert, the floods of origin inside him?*[30]
>
> - Rainer Maria Rilke

These children who cannot accept the myth have few options, none of them favorable from a societal perspective. At best, they become the artists—the painters, the poets, the writers, the sculptors, the musicians. If through their art they are able to give the public just enough reality—to touch the places of supreme beauty that people have forgotten about—then the artists are revered and honored, though they often remain poor. If these same artists take the brilliance of their talent and vision and compromise it (trashy films, commercial art, and much pornography falls into this category), giving society the kind of low-grade "art" it wants to fill the collective sense of emptiness that many of its members feel, they may become rich, though they are often left feeling unfulfilled as a result of their compromise.

Others, artists or not, will not compromise their vision. They look around and see little boys in adult bodies married to Miss America wannabe's, and are devastated by how they see their "happily married parents" treating each other, and by the overall emptiness and superficiality of their culture. So they paint it black, write it bold, or scream it loudly at their heavy metal concerts. They wear it bluntly in the piercings on their eyebrows, lips and cheeks, and on their shaved heads and spiked leather jackets. Needless to say, the average person in an "everything-is-just-fine" society does not want to face the reality of this much pain (which is really a pain that belongs to all of us, only a very few individuals take it upon themselves to express it), and so their materials are not printed, their lives not portrayed (except as derelicts), their music not heard, and their pain not acknowledged. So they scream louder.

When the pain is unbearable—when mother and father beat on them too hard, when they are unable to buffer themselves against the horrors of the surrounding society, and the depth of wounding will not allow them to adjust to the surrounding culture—the criminals, drug addicts and murderers emerge. Or, they "check out" and become another statistic on the list of the mentally ill. They become psychotic, autistic, borderline or schizophrenic. They commit suicide or develop a brain tumor at age five or fifteen (suggesting that they knew what they were up against and couldn't bear a life of fighting). The culture of abuse, seeing them as separate and "other" than itself, hides them as far away as possible in ghettos, prisons and mental hospitals—providing increasingly large amounts of money for security, while daily eliminating funding for the welfare, education and family programs that are the only possible remedy for these problems.

Still others become the visionaries—they are living experimental lives on the frontiers, offering their bodies, minds and labor to alternative possibilities. They are living in non-traditional communities, exploring spiritual practices, practicing organic farming, leading self-help seminars, studying holistic health and so forth. Some of their experiments are working, others aren't. Some are naive idealists, others are brilliant and generous people willing to risk everything for their highest ideals, and still others are dubious of widespread change, but are propelled by their integrity to refrain from participating in a destructive system.

Overall, society rejects such people (though this *is* changing, as some of the less-radical of these activities and practices are becoming trendy), as if they succeed in creating a viable alternative for the masses the status-quo economy would collapse. The government takes the failed experiments and spreads pictures of them all over the media to instill fear and prejudice in the public against these groups in order to maintain the effects of the cultural anesthesia it has injected into its people to control them. Groups such as the Cult Awareness Network were created by the wealthy and the afraid to assist parents in avoiding the pain of the family setting and culture in which they have raised their children; instead such groups point the finger at organizations whose leaders purportedly "brainwash" members into thinking that something is wrong with society as it is. Such "cults" are stereotyped instead of examined, as the non-violent lifestyle that many groups are pursuing may actually incite people to examine their own lives accordingly. Many high-principled alternative groups and communities are on the front lines of the battle for social change, and getting persecuted for it.[31] Lee Lozowick says:

> *An honest and sincere search for truth puts you in danger of being accused, abused and ridiculed by society. Sooner or later we'll all be called to stand for what we feel to be the truth.*[32]

Denial

> *...The first step to solving our problems in relating is to acknowledge that they exist. To do this we have to be willing to call into question both ourselves and the entire collective in order to discover the causes of our disturbed consciousness.*[33]
>
> – J. Konrad Stettbacher

If, as individuals, we were fully conscious of the pain that we bring to others by our ignorance of the aggression, defensiveness and pain that underlies our interactions with them...and if, as members of a nation, we understood the amount of death and destruction we participate in every day by way of our tax dollars and our refusal to act...if we were suddenly called to face this directly with no buffers—we would likely *die* of the pain. For this reason, we set up elaborate systems which serve to negate, or at least soften, the stark reality of our experience.

Is it possible for an entire culture to sustain a state of denial? Absolutely. How do soldiers go through military training in which they are taught to murder and torture other human beings, and then go into Third World countries and drop bombs on villages of thousands of innocent men, women and children? How do mothers in cultures across the globe sit and watch their husbands beat their children? How do owners of multinational fruit companies in the Third World allow of thousands of workers to inhale chemicals that will result in impotency and other life-threatening diseases?

We don't consciously acknowledge the pain of such atrocities. It is deeply buried through *denial*—a brilliant and complex defense created by the mind of the child, and then incorporated into the collective "cultural mind" to help the individual or group of individuals cope with otherwise devastating circumstances. To understand how denial works, take the example of Jerry, a young boy whose life is clouded by the sexual abuse perpetrated on him by his older brother, Jim. At least twice a week, his brother Jim crawls into bed with him, often asking various sexual favors of him and touching him in distinctly inappropriate ways. He tells Jerry that if he tells their parents what goes on between them that they will think that Jerry is a liar and will punish him. When Jerry protests his treatment, Jim deliberately tries to confuse him by telling him that he is dreaming and will not remember anything in the morning. Jerry knows what is happening, yet due to the fear of losing the love of his parents and his brother, he makes a conscious choice to block Jim's activities entirely from his mind, and by morning thinks nothing of them. If he did not deny, he would be devastated, and at the brunt of at least his brother's criticism, if not his parents' as well. Furthermore, Jerry will do anything to avoid admitting to himself that his big brother does not love him. The same mechanism is operative in Jim. Jim could not live with himself knowing how he has betrayed his brother and his parents, and so he wipes his own behavior off his conscience, perhaps even replacing it with a different story which he will come to believe. He simply turns his attention elsewhere.

But who is really making the choice to deny? ***Born on the Fourth of July,*** Oliver Stone's popular movie about the Vietnam war, opens with a powerful depiction of Tom Cruise as a young boy at a victory parade for the soldier's returning from World War II. The scene is brilliant—a sunny day, marching bands playing drums and horns, majorettes twirling batons, and young and old people alike cheering and screaming; a feeling of great victory in the air. The young boy is captivated by the glory of the parade, which makes such an impression on him that as a young man he enrolls in the army and goes to Vietnam where he fights and kills many people.

Who is to blame for the deaths of the people he kills? We could blame the young boy as it is he who is ultimately responsible for his actions. Yet he didn't grow up and think, "I want to go to a foreign country and murder people, and it is all right because they are of a different nationality." He could not, in his right conscience, think and act in this manner. He was a teenager when he went to the military training camps where minds more sophisticated than his own, brainwashed him into thinking it was O.K. to kill. We could also blame the country—the nation who celebrated the explosion of an atomic weapon on a foreign country. But the nation is simply one of many such nations involved in a "kill or be killed dynamic." We could blame the culture that allows such activity to exist, but the cultural mind is the result of centuries of conditioning. What allows such activity to transpire is denial—it keeps us alive and sane in an often impersonal world.

How would it look if we were not in denial? Shortly after beginning my career as a therapist, Keri, a twenty-one-year-old woman, came to me with issues of severe abuse by her father who was a church minister and highly respected in his upper-class community. For many months I struggled with her inability to trust me enough to let down the mask she was wearing to "hold it together." One morning we had an early session and Keri came in totally different. She was withdrawn, nearly despondent, and would not—could not—answer even simple questions. It was as if an angry corpse with a gag over its mouth was sitting before me. I left the session concerned, frustrated by my seeming inability to get through to her, and wondering if she would return to therapy.

The next week she came in, sat down, and said, "Well, I gave it to you."

"What did you give to me?" I asked her.

"That was me without the mask."

Her words provoked a stark awakening for me. I realized that the despondent young woman who had sat before me a week earlier was *not* in denial. The more I considered, however, I saw further than even this despondent one still had overlays of denial and that who Keri *was* lay still more deeply buried beneath the pain that she had briefly revealed to me.

In a similar way as Keri "broke down" into despondency when she let her mask down, contemporary society is also breaking down as it is confronted with the nuclear age, with marriage and family breakdown, with angry youth and with increasingly impersonal institutions. Evidence of this is the extraordinary number of people diagnosed with psychological illness, anxiety, depression, ulcers, high blood pressure, heart problems, cancer and so forth.

Imagine what would happen if for one week an entire country—including all of the government officials, CEO's, criminals and children—went off of anti-depressants, cigarettes, coffee, alcohol and sugar, and turned off all of its television sets and video games, and the government placed a national ban on all forms of violence within or by any group—from the home to the army.

The whole country would go berserk—literally. The first thing that would probably happen is that anger and depression would temporarily increase. All the rage that is otherwise smoked away, taken out on video games, lived out through violent television shows and movies, stuffed with sugar and drowned in alcohol, would be seething at the surface of people's skins. Yet the ban on violence would prohibit the display of it, and people would be forced to stay at home in bed writhing and sobbing. They would be cornered into feeling the pain of the emptiness of their present lives, and before that, of their childhood, and before that, their birth, and before that...they would feel themselves being shattered by this pain, piece by piece—feel their whole identity falling apart, feel like everything they had ever thought they knew was a lie, feel like they were dying...ideally they would all come out the other side, and, having felt the pain that is so feared, the pain that is considered to be so awful and undesirable, they would be left with an emptiness and spaciousness unlike the static, merciless depression they have come to think of as emptiness. From this feeling, otherwise known as compassion, many would not be able to return to their present line of work, aware of its effects on others, and those who did go back to work would be different while doing it. False and pretentious aspects of relationship would start breaking down in every home, and people would be faced with their lack of love. Life as we know it would necessarily collapse, yet something new and wondrous would necessarily arise in its place.

I am not naively suggesting that such a plan could, or would ever be carried out. Instead, I am pointing to an entire system set up by wounded individuals designed solely for the purpose of defending itself against its pain.

Just as the child numbs his skin as not to feel his parents' blows, so a culture numbs the heart of its people to the violence, rape and murder it proudly entertains them with. And, as the mother looks the other way from the abuse occurring around her, so the culture diverts our attention from the atrocities it perpetuates by boasting images of the "white-picket-fence life" in commercials, advertising, television, magazines. And, as the sexually abused child vows to keep the abuse secret when his life is threatened by the abuser, so the majority gives up its voice in fear of the consequences of being outspoken.

The deeper the underlying pain of an individual or a society, the more extreme forms of denial it will manifest, whereas when there is less pain and wounding—both personally and culturally—there is less to cover up. A man who has never been beaten, shamed or neglected by his parents will have no need for, nor tendency toward, excessive drinking, drug abuse or violence toward his wife or children (to assert his power over). He will live fully, but simply. The culture that is peaceful, harmonious and balanced will not revolve around distraction, addiction, abuse and the denial of it. Unfortunately, most of us do not know such a life.

A respect for denial begins to dawn on the individual as he or she comes to understand the devastation which it was created to cover up. A culture of abused people who are run by abusive systems cannot help but manifest as a culture of denial. Unfortunately, the cost of denial is high. For the child, it was often the most essential and vulnerable part of himself or herself that was wounded, and it is that part that went underground in order to not feel the pain. For the nation, it is the True Culture—the heart and heritage of the people—that must be sacrificed if it continues to support war, oppression, discrimination and economic injustice.

As the individual must eventually come to a point in which he chooses either to face himself honestly, or to live out a life of falsity, there comes a time when a nation and its members must ask themselves, "Is it worth it?" "Are we willing to live in a culture of superficiality, distraction, violence and depression—a culture that is fully equipped to destroy all life on the planet?" "Are we willing to sell the soul of the nation to avoid facing the damage we have already done and continue to perpetuate?" Thus far, the nation as a whole and the majority of its members are, by their actions and lifestyles, saying "Yes—we are willing to sacrifice ourselves and the heart of our country to keep from feeling this pain."

Wounds do not heal until we heal them. Wounds tend to fester and infect over time if not tended to, and thus the lives of both the individual and the collective will manifest stronger and stronger symptoms—calling our attention to the underlying wound in any way it can. Crime rates and child abuse statistics will continue to rise, depression will further cloud any remaining traces of national integrity, nuclear build-up will continue, reports of earthquakes, flooding and natural disasters will cover the front pages, and economies will collapse left and right. Eventually, we must break through our denial or die of our own stubbornness. The choice is ours.

CHAPTER 5

On Healing Through Touch

Work of the eyes is done, now
go and do heart-work
on all the images imprisoned within you; for you
overpowered them: but even now you don't know
them.[1]

- Rainer Maria Rilke

The healing science of homeopathy is based on the principle that "like cures like"—disease is healed by ingesting small amounts of the very elements that produced the original symptoms. Similarly, the same touch that was once used to inflict pain upon a person, now given with intention and in the proper dosage, is the very medicine that can heal him.

Unbonded children and adults may or may not experience complete healing of their unbonded condition.

However, healing takes place on a continuum, and there are many ways to bring an increasing sense of bondedness and wholeness into ones life. This chapter will present several "hands-on" techniques to help you heal yourself and those around you through touch. First and foremost, however, I believe that it is enough to cease to be a part of the problem of touch-starvation. If you are unable to integrate touch, intimacy and affection into your life, if all you can manage to do is to cease to perpetuate the cycle of "untouched," you are doing your part.

CEASE TO BE A PART OF THE PROBLEM

> ***Thou shall not hit, strike, degrade or otherwise harm either adult or child.***
>
> \- 11th Commandment

The way to begin a life of touch is to bring a stop to the cycle of hurting other people on all levels, from the gross to the subtle. Violence ends first on the most obvious level—the cessation of the misuse of physical and sexual touch, the refusal to engage fist fights, hitting children, slapping, etc., and from no longer allowing others in our presence to be victims of this. Next, one ceases to scream and yell at others and to refrain from all hurtful language...then manipulative and controlling actions...then sly and hurtful remarks...and on down the line. When we get down to the recognition of the finer points of abuse, we realize it's in all of us. A friend of mine who is a prominent author and devoted advocate of ending abuse on all levels suggested:

> ***When you wake up in the morning, you look in the mirror and say to yourself, "I am an abuser. I'm just***

going to try not to abuse anybody today." That's where we're starting from. By refraining from abusing one another, we are contributing to the solution.

It is a radical approach, but we can acknowledge that "front and center" is the only place to put ourselves in terms of touch. Observing closely and honestly the ways we hurt others, we can say "no" to the abuser inside of us.

A useful approach in working toward ending violence in ourselves is to notice when we are feeling grouchy, angry, or irritated, and therefore more likely to be hurtful toward those around us. Instead of obsessing on all the things that are wrong with so-and-so or such-and-such, or all the reasons we rightfully have to be angry, we can experiment instead with asking ourselves, "What am *I* in pain about?" (even if we think we are not in pain), and wait for the answer. When we point the finger inward, we cease to victimize those around us.

From the moment we make a firm decision that abuse must end in our lives, our inner and outer worlds begin to change. Eventually, we must turn our touch inward, for there is nobody else who can stop the war within us, and nothing else that can stop it save this touch, this profound understanding, this compassion. We touch ourselves by allowing ourselves to breathe deeply, to keep the company of good people, to refrain from addictions. We cease to persecute ourselves and insist upon participating in activities or ways of life that do not require self-numbing. We let ourselves feel, and cease to judge ourselves. We allow others to love us.

HEALING THE UNBONDED CHILD

You may have raised, or be in the process of raising, an unbonded child and not even know it. Bonding, however, can happen at any age under the right conditions.

It is amazing to realize that with few exceptions, every child at every age loves their parents. Even children whose parents have abused them, even children who *hate* their parents still love them! Another amazing thing about children in relationship to their parents is their tremendous willingness to forgive. No matter how old the child is, when the parent starts to act differently, the child responds. I am not saying that when the parent calls the child or writes him or her a weepy note of insight and apology that the child responds immediately—that's not acting differently. The acknowledgment of having not loved a child well is important, but a change in *action* holds the capacity to heal the wounds of childhood.

How one will show up differently with his or her child depends to a large degree on the child's age. Some proponents of shared sleeping say that if a child did not sleep with the parents in infancy and early childhood, that the missed bonding can still occur if shared sleeping starts when the child is seven or eight years old. It's a sensitive issue, certainly. Parents who feel comfortable with shared sleeping can at least open up the invitation, and let their child decide, as long as they are clear that they are equally all right with a "yes" or a "no" response.

Healing the bond with a child needn't be that drastic however. Many parents, especially fathers, are literally frozen with fear when it comes to touching their children. Parents who were not given touch in their own families think that they do not know how to touch their children. Healing can begin if one is willing to prioritize their desire to change their relationship with their child, placing this desire above and beyond their own insecurities about touching and loving him or her. We swallow our pride, acknowledge our fear, and *go for it* anyway. (An important note on this point is that a child may not immediately *want* touch at this time. He or she may be wary, angry, or just not want it for whatever reason.) The specialized psychological field of "reparenting" deals extensively with these issues.

When children are older—whether they are teenagers or mature adults—the healing is different. Although the older child still looks to the parent for love, acknowledgment, and signs of change, when a child is no longer a child and is long past the stages of bonding, the most we can do for him or her is to be honest with ourselves. Parents and children are deeply connected—even in the most uncaring of circumstances—and thus when we begin to change, it is likely that our child will feel that, even before we say anything. There are numerous remarkable stories of people having insights or psychological breakthroughs in relationship to their children, after not seeing them for years, when suddenly the child will call them up out of the blue!

Beyond this, it is useful to share with our children what we are discovering about ourselves, if we are able to do so with sincerity and clarity. Our child will appreciate our vulnerability, respond how they respond, and hopefully allow for greater intimacy in their present relationships with their own husband, wife, friends, and/or children.

HEALING THE UNBONDED ADULT

As an adult attempting to remedy your own unbonded condition, the healing is different. You will not jump into your seventy-five-year-old mother's lap, and hopefully you won't try out some regressive therapeutic process that gives you a bottle to suck on and a teddy bear to sleep with. Your healing will take place within yourself, and with the help of other caring and experienced adults who share your intention to heal. The remainder of this chapter is an overview of some of the various healing processes and techniques that I have found effective in transforming a touch-starved life into a life full of touch, intimacy and wholeness.

Getting in the Body

The best way to heal any touch-wound is by "getting in" your body. When *you* (as opposed to your wardrobe, your ideas of who you think you should be, etc.) are the full-time resident of your body, then you *are* touch, and you have nothing left to do except stay that way. Unfortunately, most people in the contemporary Western world are disassociated from their bodies, yet another symptom of a culture of abuse. (See Chapter Four)

The modern body tends to be developed in one of two ways—one is the "fast food" body. The "fast food" body is flabby and often has high blood pressure or high levels of cholesterol as the result of a diet that is laden with fat. The skin tends to be dull and "washed out," having not been fed the nourishment necessary for vitality. The person may be sluggish, lazy, and tend to spend their free time lying around in front of the television set munching on a bag of chips.

The other type of modern body is the "hard body," represented by pencil-thin women with tight bellies and muscular arms, who live on Diet Coke™, Lean Cuisine™, Nutrasweet™, fat-free ice-cream and Mademoiselle; and by men with sculpted stomachs and bulging biceps whose hours off work are spent at the gym (though the illness is not as grave in men). While there is nothing wrong with being physically fit, being able to bench-press 500 pounds has nothing to do with being "in" one's body.

So, what does it mean to get in the body? That's certainly what I wanted to know when I first started hearing about this idea. People who are in their bodies—no matter the shape or size—have a certain radiance about them. They are present—here! We are aware of being listened to when we speak with them. They seem to really show up wherever they go—not just as another body, but as a part of what is happening. They are energetic, attentive, involved. They feel the movement of their breath, their feelings and how they are affected by their environment.

When we are "in" touch with our bodies, we cannot help but notice the larger body—the earth—and what is happening to our environment. We start to pay attention to the kinds of chemicals in our foods, the quality (or lack of quality) of the air we are breathing and the water we are drinking. We are unimpressed by the concepts of silicon breast implants, face lifts, and cellulite treatments. Beyond this, when we are "in" our bodies we receive a continual flow of feedback from our environment. When we look at a menu we know immediately what food would be best for us; we are able to sense danger before it occurs, and are aware of when we are in dangerous company. Being "in" the body enlivens and animates the senses—we can smell lavender bushes from a distance, taste the sweetness of a carrot, or feel our whole body relaxing in response to the touch of a smile.

When I walked in the Costa Rican forests, I was always surprised when a friend would nudge me and point to a small, but poisonous snake hiding under a leaf ten feet in front of us, or draw my attention to a toucan sitting fifty feet away high in a treetop, or predict a coming rainstorm by looking at a cloudless sky. No prior thought or analysis, just instinct. The bodies of these native people (as opposed to their intellects) had been the source of their knowledge all their lives, and was sustained directly from the earth. All their senses were strong, alert, healthy—connected with the surrounding environment.

A further benefit of being "in" the body is that it allows us to gain access to a deeper level of feeling. Feeling is registered in the body. When we feel loving, there may be a tingle in the chest, an overall feeling of relaxation in our muscles, and a slow, relaxed pattern of breathing; whereas when we feel angry there may be a tightness in the jaw, tension in the chest, and a constriction or holding in the gut. Some people live in such an emotionally disturbed state all the time that their bodies always feel this constriction and tension. Hence we see chronic back problems, ulcers, and asthma.

Twenty Ways to Get into Your Body

Since the tangent point of touch is the skin, physical contact as well as exercise and movement can help bring someone who is numb or dissociated back into their body. When someone is feeling spaced out, disoriented, or disconnected, a stroke on the hand or the arm, or a touch on the back, can bring her back to life. When I practiced as a massage therapist, clients would often come in "out" of their bodies, stressed out, and begin nervously chatting away as soon as they got onto the massage table. More often than not, as soon as I began to rub their feet or their back, their breathing pattern would immediately open up and they would start to relax.

The week after I suggested to a client that she do something nurturing in order to acknowledge and care for her body, she reported that every morning upon waking, she would spend a couple of minutes rubbing lotion on her feet, giving herself a simple foot massage. Many women spend hours putting make-up on, but my client's intention was to care for herself and honor her body. Since the feet contain very important acupressure points that are connected to all the organs in the body, massaging them helped her to nurture herself, and thus the exercise proved tremendously beneficial.

The following suggestions will immediately help bring you back into the body, though you may need to do them regularly to create lasting effects. They work!

- Weed the garden
- Play Frisbee
- Jump up and down
- Take a hot bath
- Stop and breathe deeply for fifteen seconds
- Ask a friend to give you a hug
- Play with a child
- Get a five minute back rub (or give one)
- Scrub the bathroom
- Stick your head in a pillow and scream
- Walk barefoot outdoors
- Jump on your bed
- Sing a silly song at the top of your lungs
- Run to the end of the block and back as fast as you can without stopping
- Make funny faces in the mirror
- Cry
- Put on some great music and dance wildly
- Walk or jog in the rain
- Take off all your clothes and sunbathe in a hidden spot
- Take a cold shower or plunge into a pool

Touch for Health

I was going in the hospital for my fourth spinal surgery and there was no guarantee I was going to come out of it alive, began Jake, a 52-year-old lawyer. *Although they never let spouses into the recovery room, I asked if they would make an exception in this circumstance and allow my wife, Susie, to come in. I knew what I needed...I had warned Susie that I was going to look terrible. I didn't even know if she'd want to touch me. But both Susie and the hospital staff agreed to my request. When I came out of the operating room, I was in a coma for days. Although I don't remember anything she said to me, I remember the distinct feeling of her touch. At the point where her hand touched my arm I could feel a sensation of warmth extending throughout my whole body. That warmth—the tangible feeling of being loved—is what carried me through.*

Many people actually become sick because they are touch-starved. Unconsciously seeking a similar type of attention and healing, their pain comes out in the form of physical illness. The memory of mother rubbing their forehead and bringing orange juice to them while they were sick in bed may be among their fondest memories of childhood.

Studies performed at the University of Miami Medical School's Touch Research Institute, indicate that physical touch reduces stress. Yet a 1993 U.S. Public Health survey estimated that 70%—80% of Americans who visit conventional physicians suffer from a stress-related disorder.[2] Modern medicine generally tends to treat the symptom and not the disease. Therefore, even if the individual is fortunate enough to have his or her symptoms cured, if the underlying causes, including touch-starvation, have not been addressed, then the disease will manifest sooner or later in another form.

Dr. Theresa Crenshaw, author of *The Alchemy of Love and Lust*, explains that touch alters the chemical composition of your body. "Lack of touch is just as detrimental to our health as a lack of Vitamin C," according to Crenshaw.[3] She tells us that the peptide oxytocin (a hormone-like chain of molecules that is naturally produced by the body) increases when an individual receives touch or nurturing physical contact. As oxytocin increases, the individual craves more and more touch. However, this touch "addiction" is far less hazardous than most common addictions, according to Crenshaw, as the estrogen-dependent chemical oxytocin relaxes the individual, promotes touch, encourages bonding, triggers milk let-down during breastfeeding and sets off the uterine contractions that accompany childbirth and orgasm.

Genuine and simple touch given in a safe and nurturing environment in which one can be held and allowed to feel the depth of the pain that he or she once endured, and continues to live with, is one of the best remedies available. Ideally such touch should be given by those who have been willing to explore their own pain, and who will not become frozen or uncomfortable when the other's pain is revealed. Receiving this type of touch allows one to learn to trust, over time, in environments that are loving and non-abusive—especially if the childhood environment was otherwise. While no other person can ever provide the "perfect" love that is longed for, and those helping us have been wounded too, it is still possible to receive nurturing touch from someone who is nonetheless trustworthy because of their commitment to us and their understanding of themselves.

Who are these people who can provide us with healing touch, and in what environments? Ideally, the "toucher" would be a husband, wife or trusted friend. A knowledgeable massage therapist may be the best alternative for many people. The most optimal environment is your own safe home, but might instead be a women's or men's support group. Eventually, this safe environment will be discovered in more of your surroundings—for as you heal and are able to give and receive love and nurturing touch, you will effortlessly transform your surrounding circumstances.

Your life will eventually change with the positive, ongoing benefits of healing touch, but it will take time—there is simply no instant way to affect such transformation. But isn't it worth it? You have suffered this wound for many years; wouldn't it be worth it to invest the next several years in healing it, so that you may come to live at least part of your life in the experience of love, touch, and intimacy? Wouldn't it be nice to know that you are capable of radically transforming your own pain into love? What price could be too high for that?

Counseling, massage, therapeutic touch, seeking the company of wise elders, developing contact boundaries, and initiating your own sexual healing are a few ways to treat the touch-wound outside of the doctor's office.

Holdings

Some adults approach healing through touch by doing "holdings" with one another. They make an agreement with a close friend to spend a specific amount of time together (ranging from twenty minutes to an hour, agreed upon in advanced), in which the friend will just hold them. This works best when there is a "holder" and a "holdee." In other words, one person is exclusively focused on holding, nurturing and paying attention to the other, and the "holdee" attempts to simply allow themselves to relax and be held, without having to "do" anything, just as they wished their parents had done for them in childhood.

(Holder and holdee will often alternate turns over a period of days.)

Holdings are not a time for talking and analyzing, though a few words won't hurt anything. They are what the name suggests—a time to be held. Many will find themselves uncomfortable, self-conscious or numb at first. Or their mind will wander, they will feel silly, or suddenly feel like they don't want to be there. Others will find that deep grief, tears and memories quickly arise. All of these are symptoms of the unbonded adult. Those who engage in holdings should realize that there is nothing wrong or right about any response, and no need to react or judge any emotional state that arises. Emotions and memories can just be noticed or felt. When we allow emotions to be processed through to completion, we will discover that feelings *do* change.

Counseling

Although it rarely involves direct touch, another path that people explore in their search toward wholeness is psychotherapy or counseling. In their minds, they may be turning to the therapist for help with their intimate relationship, or with personal problems at work, yet unconsciously many of these people are searching for someone to bond with in a genuine way—someone who is willing to listen to them and accept them just as they are. Many of my clients have raved about how healing their therapy was, when in essence all I did was sit there, with as much presence and as little judgment as possible, and listen to them, giving occasional feedback. Yet it is precisely this that so many children never received—no one listened to them, accepted them as who they were, and supported them in following their own dreams and aspirations.

Therapy is not a replacement for unbondedness, but it does serve to turn the client in the direction of self-healing. We are often blinded to the ways that we perpetuate our own woundedness, the ways that we block our own healing and deny ourselves affection. A good therapist can point out our blind spots to us and guide us in the right direction.[4]

Massage

Massage is another excellent avenue by which to seek out the benefits of touch. Several years ago, having just moved to a small village in England and looking to set up a local massage practice, I decided to test the waters at the popular Sunday morning flea market in the town square by setting up a chair beside which I had printed a large sign that read, *"Free! Ten-Minute Therapeutic Massage."* The responses were fascinating. Besides being confronted with peoples' extreme wariness of anybody who was going to give anything away for free (this insight alone made the day worthwhile), I was exposed to a wider range of responses concerning peoples' ideas about touch and massage than I ever had while giving therapeutic massage in a private-practice setting. "Why should I let you touch me?" or, "What's in it for you?" some asked accusingly. Several others, who had never been able to afford a professional massage, or who had been fearful, but secretly curious about it, sat down and allowed themselves to have this experience for the first time. Through a simple ten minute massage, many people were able to experience the nurturing qualities of touch that they had rarely known—I saw this when they got up from the chair relaxed, refreshed and happy.

The physical healing powers of touch—transmitted through the means of massage—allow for physical and chemical alterations (e.g., lower levels of cortisol and norepinephrine, which are anxiety and stress hormones) to occur in the body resulting not only in the prevention of illness and improved health, but in an overall sense of aliveness and well-being.

The benefits of massage are far-reaching. It is widely known that massage can heal sore and aching muscles, and release tension and anxiety, thus providing greater relaxation. But some of the other physical healing properties of massage are less widely recognized. Among adults, massage may aid in decreasing depression and lowering high blood pressure, as well as assist in the relief of symptoms associated with mental illness, stomach pain, spasms, heart disease and dysmenorrhea. It can help to restore movement in strained, fractured and wounded limbs, and it generally serves to improve circulation, help eliminate waste and reduce swelling.[5] In adults with chronic fatigue syndrome, which is believed to be related to depression, massage has proved to significantly reduce emotional stress and somatic symptoms, as well as depression and difficulty with sleeping.[6]

Children like massage too. One evening after a neighborhood barbecue I was sitting outside on the lawn when an eight-year-old friend came and sat on my lap and asked me to warm her up. After rubbing her back and arms briskly, I began to massage her body. At first I was working timidly and lightly, as most of us imagine that children's bodies are more fragile than adult bodies and do not crave the same kind of touch. However, it soon became clear that the more I connected with her as a human being, the more her body responded naturally to the touch. As we sat there under the night sky, I continued to work limb by limb, finger by finger, ear by ear, toe by toe. I was well aware that I was receiving as much from the massage as she was, as is always the case in a good massage. At one point she commented, "You know, my Mom does this too, but she's afraid of hurting me so she doesn't press hard enough to make it feel good."

Massage has been shown to improve health in diabetic children, and to decrease anxiety and depression in children with Post Traumatic Stress Disorder.[7] In children and adolescent psychiatric patients, massage decreased depression, anxiety and stress hormone levels. Massage also improved their clinical progress.[8]

The elderly, who represent the most untouched population in the Western world, thrive on massage. Many of our elderly are widowed, and unless they have frequent contact with their grandchildren or have children who are aware of how much touch an older person needs, they have little or no means by which to be in physical contact with others. Seeing this in my ninety-one-year-old great aunt, I approached her shyly one afternoon and asked if she would allow me to massage her. She laughed embarrassedly, feeling flattered and disbelieving that I would wish to touch her aged and frail body, but agreed to a massage the following day. Given her character and her initial discomfort, I was amazed to feel and watch how her body drank up the contact like a desert plant drinks up moisture. In contrast to the child mentioned above, I worked very softly and gently with her bones and cartilage, paying careful attention to her body for any signs of feedback that she may have been too timid to speak of. Geriatric massage is a field of growing popularity, but still largely untapped as a resource for the elderly.

Lastly, massage is increasingly being incorporated into the field of preventive health care. Perhaps the most inspirational of the new possibilities for massage come from studies with AIDS patients carried forth by medical doctors and students at the Touch Research Institute in Miami, Florida. HIV-positive men were given forty-five-minute massages five times per week for a one month period. The results strongly suggest that massage has an effect on the improvement of the immune system, not only with AIDS patients but in general immune function as well. In addition to improved immunity, AIDS patients also showed a reduction in anxiety, stress, and distress levels—a type of relief much needed in a dying population.[9]

You don't have to be a massage therapist to give a good massage. Anybody (yes, even you) can do it! I remember agreeing to trade massages with a man who boasted his extensive training and abilities in massage. It was terrible! His touch was technically perfected, but was totally void of qualities of presence, nurturing and care. On the other hand, my favorite massages are given both by a womanfriend who insists she doesn't know what she's doing, and also by a friend's five-year-old daughter!

Therapeutic Touch

Therapeutic Touch (TT) is another popular practice through which both professionals and laypersons have learned to heal others through their hands. A contemporary interpretation of several ancient healing practices, Therapeutic Touch is based on the premise that it is the patient who heals himself, with the healer acting as a conduit for the necessary healing "energy." A form of healing that focuses on balancing the energy field of whole person rather than treating specific physical diseases, Therapeutic Touch usually does not involve any physical contact; instead, the healer's hands are placed about two inches above the patient's body. The main benefits of

Therapeutic Touch can be divided into four categories: relaxation, pain reduction, accelerated healing and the alleviation of psychosomatic symptoms.

A non-traditional method of healing, Therapeutic Touch has earned its place in the field of traditional allopathic medicine. In 1975, co-founder Dr. Dolores Krieger published a landmark study in the *American Journal of Nursing,* documenting a significant increase in the mean hemoglobin level of a group of patients in New York City hospitals who received Therapeutic Touch over those who received routine nursing care.[10] Since that time, it has been taught to more than 70,000 nurses, in more than eighty colleges and universities in over seventy countries.

Seeking the Company of Wise Elders

Seeking the company of wise elders is an invaluable source of affection and nourishment that has been largely forgotten in the West. Wise "elders" do not have to be old, but they do need to be wise. They are individuals whose lives are an example of touch and intimacy, whom you can look to for guidance in your own life.

One of my favorite people to hang out with was my grandfather. There he was, a ninety-two-year-old, Republican businessman with highly conservative mannerisms and ideals—a far cry from anyone I would ever have expected myself to seek out. Yet, to be in his company was to be in an environment of pure and unconditional love and acceptance. He had moved beyond critical judgmentalism and binding attachments and into an appreciation and detached observation of his environment. Sometimes I could not even think of anything to say to him, but I would stay in his presence nonetheless, just to feel his love.

Unbonded adults can move toward greater touch and intimacy in their lives by spending as much time as possible in the company of those who are bonded. The idea here is that intimacy is contagious, and that when you spend enough time in the company of intimacy that you "catch it." Everybody has probably had the experience of being in a bad mood when suddenly greeted with warmth and tenderness by a mate or friend. Instantaneously you find yourself in that same mood, and responding in kind. As you spend increasing amounts of time in the presence of intimacy, you allow your true nature to come forth—the innocence and openness that was there before the period of missed bonding. Simultaneously, you are reprogramming the false messages that you may have received as a child that intimacy was dangerous, and in doing so you open the door to deeper contact.

When I initially decided that I would try to "catch" intimacy, the first question that came up was, "How can I tell who has this intimacy?" It's a good question, and I quickly understood that discovering who these people were was a substantial part of the exercise. Those who were not bonded as children are not attuned to what intimacy looks and feels like. Their attention is usually tied up in unconsciously attempting to defend themselves from letting anybody get too near to them while at the same time they grasp at any love that they can find. They have not really considered what intimacy is, nor have they aspired toward it. Therefore, as I began work on this intimacy hunt, I began to look at the people around me—to watch how they interacted with their husbands, wives, children and friends. I began to watch who touched one another and when, to listen to their tone of voice, and to watch the subtleties in their actions and movements. For people who have spent their lives unknowingly hiding from intimacy, it can be fascinating to watch the way others relate.

One of the first things I noticed was that there weren't so many people who had an ease with intimacy, in spite of superficial appearances. There were situations in large groups in which I would look around the room and estimate that there were probably no individuals in that room who had properly bonded as infants. Nonetheless, there were clearly some people who had more facility in relationship than others, and thus I began to keep their company.

Seeking the company of wise elders, or those for whom intimacy is natural, is an ongoing endeavor. You won't find immediate and measurable results, even if you *are* being infected.

Developing Boundaries

Developing clear personal boundaries will not only bring more caring touch into your life, but will provide you with a sense of clarity and certitude about yourself. Boundaries are a hot topic in pop-psychology. People are running around left and right setting supposed "boundaries:" "I have to set a boundary that you must refer to me as a woman and not a girl." "I am making a boundary that you cannot put your arm around my shoulder unless I give you permission to." When people say, "I know how to set a boundary," what they often mean is, "I read in a pop-psychology book that I can control a situation by saying, 'I need to make a boundary.'" Although a step in the right direction, these types of boundaries are usually rigid and superficial. A person who is clear about boundaries within herself would never need to use the words, "I have to set a boundary." She is able to assert herself clearly and directly, but at the same time in a manner that is neither aggressive nor defensive.

Real boundaries come from the inside. They come from knowing where we end and another person begins, from knowing who we are and what we want, and from understanding the signals of our bodies. When we know ourselves from the inside out, we can create or relinquish boundaries at will depending upon the situation. For example, at a bar or on a crowded subway, one can, through the power of his or her attention, energetically place a boundary outside of themselves that will not be permeated by any "dark" forces around them. On the other hand, in a sexual relationship, a person who knows himself well can intentionally allow his boundaries to dissipate.

Good boundaries around touch come from knowing your own motivations. They emerge from a conviction that says, "Here, in this circumstance, I know myself well enough to touch you." You are aware of why you are extending yourself to the other—be it as a gesture of friendship, reaching out from loneliness, a desire for pleasure, or a wish to comfort. Any of these are fine motivations for touch if your intention is clear. However, when your motivation to touch someone is one thing, and your actions suggests another, you will confuse people and are likely to feel dissatisfied by their response to you.

When you are "in" your body, and are able to express and receive touch in a healthy, consistent manner, boundaries fall into place naturally. When you have clear boundaries inside yourself, it is unlikely that others will shy away from your touch. You will instinctively know who to touch and who not to, as well as when and how to touch them. When your judgment errs, you do not fall apart or feel ashamed for being inappropriate, nor do you blame the other for their frigidity. You just bring your touch back into yourself and go about your day.

If you are a man or woman who tends to shy away from touch, you can experiment with boundaries by deliberately being more affectionate. On the other hand, if you consider yourself, or are considered by others, to be a "touchy-feely" person, you may want to experiment with refraining from touching others for a day or two, to see what happens. Bring uncompromising honesty and attention to who, when, and why you touch (many "touchy-feely" people are actually terrified of intimacy, and use their hands to hide their fears—even to themselves).

People who have been abused, or those who have a difficult time touching and being touched by others for any reason at all, not only need to re-own their desire for touch, but to re-sensitize themselves to it. They need to make distinctions between a touch-deprived state of panic that will cause them to indiscriminately allow anyone to touch them in the hope of filling the hole of emptiness within them, and a desire to be touched based on a feeling of fullness and affection. In children who have not been abused, this ability is so natural that it is not, and need not be, conscious. But when people who have been abused are faced with opportunities to be touched, their unconscious is often so flooded with feelings of anger, self-hatred, overwhelm and fear, that the desire for physical touch cannot be distinguished from these other, older feelings.

Contact Boundaries: An Exercise

The following exercise on contact boundaries has been helpful, and even fun, for many of my clients. The purpose of it is to re-learn to discriminate and to express what kind of contact feels comfortable for you.

Experiment with a trusted friend, counselor, or spouse in learning your contact boundaries, making a clear agreement in advanced that there will be no sexual contact. Generally it is best to experiment with someone of the same gender. Start the exercise standing across the room from one another and begin to walk toward each other. You may ask your partner to stop at any time. Pay close attention to what you are feeling as you move closer. You may be concerned about the other person—that you will offend or insult him or her by either your nearness or your distance. Realize that this is part of what comes up for *you* around touch, so take it in as information, while you maintain focus on yourself. At some point you will literally walk smack into an invisible boundary. Stop there and take note of it. Look at your friend. Move slightly backward and forward to explore the immediate parameters of your boundary.

If you want to explore this further, take a step past your boundary and let the thoughts and images that arise be there. Stand in the discomfort of going past your comfort zone and allow it to inform you. You may want to hold your friend's hand. You may want to drop it...then pick it up again. Be creative and have fun with it if you want. There is no right or wrong, good or bad, there is only observation and experimentation. (Contact exercises in relationship to sexuality are discussed next section, "Sexual Healing.")

This exercise is not intended to locate a single and fixed contact boundary. We are not looking for something static. Our relationship to life is always in flux, and boundaries become rigid when they are defined and made into rules. The exercise is practice in feeling your own body in relationship to touch and contact—to learn what you want and what you don't want, and when. When this becomes clear, you may still choose to have contact even when you don't necessarily want it (e.g., parents who are available to their children whenever the children want affection irrespective of their own needs), but you can distinguish your own needs from those of others, thus taking responsibility for whatever choices you make.

Sexual Healing

A further implication of contact boundaries applies to individuals who have been sexually abused, either subtly or overtly. Those whose sexuality has been used to harm them will often either be terrified of, or addicted to pleasure. Regardless, their sexual boundaries have been stripped from them and until they re-educate their bodies, they will not know their natural sexual rhythms and cycles. Instead, their bodies will be at the effect of rigidly imposed mental constructs about what is safe in terms of sexuality.

Similar to the above exercise on contact boundaries, therapists who specialize in the field of sexual healing will consistently recommend the following practice for couples. The purpose is to assist them in exploring their bodies' physical responses to touch, free from sexual pressure and habitual mental responses.

- *Exercise:*

The couple comes together with a shared intention to help one another become more aware of their own bodies, deciding ahead of time that there won't be orgasm, and that there will be no pressure to feel anything in particular, nor to provide pleasure for the other. Alternating between a focus on giving and receiving, the couple then experiments with any form of sexual or non-sexual touch they wish to—learning to speak freely about what they are feeling, and asking to be touched in one way and to stop being touched in another way. They allow themselves to feel pleasure, pain, rage, numbness—whatever is present. Sometimes one of the individuals will simply want to be held. The point is to open an emotional/psychic space within the couple that is wide enough and safe enough to feel all things—a space that allows each person to experience in the present how he or she is impacted by the other person's touch. Given the safety to feel, free of expectation, an individual will often discover that there is a significant difference between their past experiences of sexual contact and their present one.

It is particularly important in this exercise to allow either person to withdraw from contact. If the couple was engaging in an ordinary and unintentional sexual experience, withdrawal from contact might result in resentment, confusion and hurt feelings; however, in this exercise, each individual can explore both their desire for contact and their wish to withdraw. Many people who are disconnected from their bodies or sexually frozen do not believe that they have the right or ability to exercise choice in terms of being touched or not. The beauty of this exercise is that when one allows himself or herself the liberty to withdraw from contact, they simultaneously gives themselves the latitude to engage in sexual contact and to long for it. The freedom to feel all things includes the freedom to want and to desire, unweighted by pressure and expectation. This exercise need not be serious and planned out in advance. Many couples use it as a form of play, bringing humor to otherwise uncomfortable dynamics and enjoying the interchange between them.

Learning to feel one's own body in this way, to discriminate between pleasure and pain, and to communicate clearly about one's own sexual preferences are important stages of healing, and cannot be passed over prematurely. However, there does come a time when the need to heal one's own sexuality has passed, and the person can consider placing his lover's sexual preferences and contact needs over his or her own—to touch the other in a particular way simply because that is what he or she wants, and because there is pleasure in the giving. This is where all healing leads to.

CHAPTER 6

Toward a Healthy Model of Sexuality

In the Western world it is highly probable that sexual activity, indeed the frenetic preoccupation with sex that characterizes Western culture, is in many cases not the expression of a sexual interest at all but rather a search for the satisfaction of the need for contact.[1]

- Ashley Montagu

For many people, there is a great deal of confusion and difficulty in the area of sexuality. We live in a culture of promiscuity where there is an obsession with sexual icons, phallic symbols and body image. Sexual dysfunction is more common than not, and the popularity of sex therapy has soared in the past decade. In a touch-starved nation, many people were badly abused, and few individuals were shown healthy models of affection in their homes. Whether this cultural illness manifests as an obsession with sex, an aversion to it, or a fear of it, it is no wonder that so many people in a touch-starved nation are concerned about their sexuality.

Understanding that the roots of sexual dysfunction are less a consequence of personal neurosis, and more the result of a culture that has lost touch with the sacredness of the body and of sexuality, we can start to look for a new *context* from which to base our approach to sexuality. We need to recognize the personal and cultural biases, expectations and ideas that inform our approach to sexuality, and learn to discriminate between those perspectives that are subjective and debilitating, and those that are conscious and healthy.

We learn discrimination in the area of sexuality through a process of experimentation, and that means that at times our experiments will be unsuccessful. Nonetheless, each gesture we make toward developing a healthy model of sexuality in our own lives is not only a step toward making our own sexuality more conscious and fulfilling, but is also a small contribution toward eradicating the disease of sexual dysfunction in a touch-starved nation.

12 HINTS FOR A HEALTHY SEXUALITY

Hint #1—Sometimes Sex is Just Sex

Sometimes sex is just sex. You just do what you do, as long as it is agreeable and enjoyable to your partner, and it's totally fine. Intimacy doesn't always have to be work. Sometimes you just want to be raw and mindless, and it doesn't have to be a problem. It can be more problematical to make grandiose efforts to have profound and meaningful sex just because that is what you think you should be doing.

There *is* a time and place for "higher" sexual values, but sometimes sex is just sex. You act as though you were teenagers in the back seat of a car, or you roll around on the ground like animals. Animals don't verbally process their whole sexual encounter before, during and after the act. They just jump on top of each other and do their thing—and sometimes that's what's needed.

People's minds are so full of ideas about how sex is supposed to be that many people are afraid to even get in bed. Some people believe that if they aren't master sex technicians that they will not be able to satisfy their partner. Others think that sex has to be a major cosmic production every time, weighing themselves and their partners down with unreasonable sexual expectations that no human being can live up to.

On the other hand, if raw, mindless screwing is all your sex life is about, you're missing out. Sometimes sex is communion, sometimes it is love, sometimes it is a union, an exchange, a deepening, a bringing together. Sex is sex and love is love, and sometimes the two mix and sometimes they don't. There can be love in the relationship and still have sex be sex sometimes. Or there can be sex with no love, which over the long run will not be ultimately fulfilling. When we allow ourselves to relax a little bit, letting sex just be sex sometimes, we make more room for sexuality to develop fully, organically, and free from the confines of our expectations.

Hint #2— Face the Facts: Where Sex is Concerned We're All in the Dark

> *Why are we so into sex? ...because we die somewhere in it. We die in sex, and that's what we're looking for. We are desperately looking for something to make*

> *us forget this disruptive world we live in, and sex is supposed to make us forget. We keep using it more and more to that end. But, it doesn't really work. We want to die desperately, to end our suffering. We want to be born again desperately, to feel. We want to wake up desperately. That's the big attraction to sex.*[2]
>
> – LeeLozowick

We currently live in the dark ages in terms of sex. I don't know what it was like two-hundred years ago, but I know that we as a people are becoming increasingly ignorant in our knowledge of sexuality. Let's face it, when sexual abuse statistics are documented at as high as fifty percent, something has gone haywire. Sex has gotten twisted up with affection, aggression, desire and love.

Sex therapy has become such a big field in the counseling profession because people need help. The sex therapists I have spoken with say that everybody comes in with different variations of the same drama in terms of sexual problems, but that the root of the problem has to do with the quality of the intimate relationship itself. Some sex therapists say that years of impotence can quickly dissipate with a clear enough insight into the problem. Sexual problems usually stem from a blockage in the flow of intimacy between two partners. After the first two weeks (if you're lucky), two months, or year together, sex develops into an expression of your relationship with your partner. If either you or your partner are secretly angry, jealous or resentful of the other, those feelings are going to come out in your sex life.

The point is that we don't have a deep understanding of sexuality—either our own or our partner's. We're generally in the dark as far as sex is concerned, and it's not just that we have the lights off. As men and women, we have so little understanding of one another and are embarrassed to communicate about our most basic sexual needs and desires. We are like a bunch of sixteen-year-olds with wrinkles and graying hair trying to act like we've learned something about sex.

Why are we so in the dark? Because the situation is more complex than knowing how the genitals work and what it feels like to get horny. The problem has much more to do with touch-starvation, touch-phobia, touch miseducation, and an out-of-touch society that has left everybody ignorant concerning the most basic principles of touch.

Beginning with the family, there is a gaping void in our culture when it comes to sex education. If children learn about sex through a story about a bird, they are going to know that their parents are greatly insecure about the topic. If they learn about sex from asking their big brother or sister about it, or from stealing their father's *Playboy* from under his bed, they're not only going to get the wrong idea, but will be left highly suspicious of what sex is, and why their parents didn't tell them about what they are seeing in the magazines.

And it's no better in the schools. When I was in eighth grade, our sex-ed teacher—whom we secretly called "Dr. Ruth"—taught us how to put condoms on wooden penises, and had us sing songs about penises and vaginas so we could get over our fear of speaking the dreaded words. She taught us about venereal disease, but not once did she mention the sacred aspect of sexuality. Nowadays, the first thing they teach school children about in sex-ed is AIDS. And they should, because the concern is very real, but it leaves children with the idea that sexual touch is a lethal pill.

A society that is in the dark in terms of sexuality will lack tolerance for those members who want to live according to a different model of sexuality. Orthodox Jew Elisha Hack, a twenty-year-old freshman at Yale University, opposed the lack of sexual modesty at his college—men and women living so closely together, stashes of condoms laying out in the open. Yet when he proposed to live off-campus in order to be aligned with his Orthodox values of sexual modesty, his request was refused.[3] The decision-making body who denied his request found it more important to comply with rigid administrative regulations than to honor this man's desire to adhere to a more modest form of sexuality.

On the level of culture, the closest we get to sexual education are articles in fashion magazines with titles such as, "The 7 Little Signs He That Loves You When He's Too Chicken to Say It," "Sexual Confidence—Bold Moves to Make Tonight," and "What Makes a Woman *Bed*able? Men Reveal the Surprising Things that Make Them Lust After You." On top of that, the television sells countless images of an impossible sexuality—passionate fifteen-second blips that never even occur in real life. There's nothing more annoying in bed than a man who is convinced he is James Dean or a woman who insists she's Madonna—only better. The media has sensationalized, improved, distorted, and cut and pasted ordinary people's clumsy sexual encounters, not to mention their bodies, in such a way that everybody is left thinking they have to be some superstar sexual performer.

I am always shocked by all of the mature, supposedly responsible men I know who will fail to use a condom when they have sex with a woman whom they hardly know. Besides the fact that these men themselves could wind up with venereal disease, warts, or worse, I always ask them, "Are you willing to be a father of this woman's baby?" They routinely answer something like, "Well, she didn't say anything about it so I'm sure she's taking care of it." I tell them they've gotten stuck in the dark ages. There are foolish and touch-starved women out there who are going to sleep with a man without taking care of themselves just to get a little bit of love. It is up to the man to take equal responsibility.

In this culture, sex is the most available and widely acceptable touch pill. We crave contact. We want something to fill the hole of emptiness. We also sense a higher possibility (yes, even higher than a five- or ten- or even twenty-second orgasm). We want to merge, to dissolve. We want a little piece of that pie that some call God, some call The Mystery, or some just call bliss. We know of few other doors to freedom so we knock on the one that says SEX. And it is a good door, for it *can* take you there. But it will never take you there if you fake it, or try to moan like you're there when you're not, or if you do and say things just because you've read in a book that it worked for somebody else.

In terms of sex, we've got to start from where we are, because if we're trying too hard to be somebody we're not, we're going to say or do the wrong thing. Furthermore, we're bound to make mistakes because we are human beings, and even if we've participated in the same sex hundreds or thousands of times, sex is real—it is the body, it is vulnerability, it borders on the mystical and cannot be controlled.

Healthy sexuality starts with an admission that most of us don't have a clue to what it is. To get out of the dark where sex is concerned, you have to turn the lights on inside. It can be quite freeing to let all the skeletons out of the closet—you may just find that hidden underneath all those old bones is a great treasure, and something real.

Hint #3—Get in Bed with your Lover—Not Mommy and Daddy

> *...The confusion between the need for sex and the need for affection, a maternal kind of physical contact...is important to resolve. I believe that with a clear notion of the distinction, and a little practice in dissociating the two, a great deal more affection could be exchanged without the complications of sexual involvement when it is not wanted.*[4]
>
> - Jean Leidloff

Mommy and Daddy have this habit of invisibly creeping under the bedcovers with us and screwing everything up. Though many parents were excellent role models of loving relationships, instead of sleeping with our lover, most people jump in bed with Mommy and Daddy, and that's why the bed is so crowded with expectations and ideals. This is not to be confused with the Freudian perspective about the child's latent desire to sleep with his or her parent. Instead, when we bring Mommy and Daddy to bed with us it is because we unconsciously perceive sex as the arena in which we can possibly fulfill some of that deep longing for touch, affection and love that is present in everyone, but may have been lacking in our childhood.

Sex is the place that we curl up to "husband-Daddy" and unconsciously say, "Please give me the attention and affection that my distant and over-worked father never gave me." Or we unknowingly say to Mommy-girlfriend, "I've been hungry for affection for the past twenty years—please fill up that hole!" Until we finally get it straight in our own minds that every lover *is not* our mother or father, we will constantly do a "mother/father number" on our partners—whether we try to gain their approval, rebel against their control, or argue to get our point across.

We crave attention. Women are not dressing themselves in clothes that leave them almost naked, and men are not "pumping iron" in the gym all day long as an expression of self-assuredness and ease within their bodies—they want attention! The men whom the girls want to attract, or the women whom the guys hope to "score" with, are really just mother/father substitutes. We scream at our idols: "Look Mom, look Dad—aren't I beautiful, see me!" The opposite extreme are beautiful men and women who, having received too much of the wrong kind of touch in childhood, hide themselves behind barriers of too-large clothes, excessive weight, and poor hygiene.

You'll also find this "Mommy/Daddy Projection" among some zealous feminists who feel the need to punish the entire patriarchy by taking out their rage on every man they sleep with. But placing the whole burden of one's childhood on one man's shoulders is just too much weight to ask him to bear. There is no way that one man can make up for ten thousand years of cultural conditioning!

Other women are consciously aware that sex is the price they pay for being touched and held. One of my clients had the following insight:

> *I became a seductress because I wanted to feel loved. If I wasn't going to be loved for myself, then I wanted to be loved for my intelligence; if not that then for my looks; if not that then for my sex.*

Particularly among men, sex has become one of the only socially acceptable avenues by which to attempt to fulfill their desire for intimacy. Oftentimes, when men are aggressively seducing or pressuring a woman, the actual need is that of a little boy saying, "Mommy, I am just dying to be held and to feel loved by you." Generally speaking, when a man needs touch, he automatically assumes that his only option is sex. Whenever he feels something that might be called "warm" or "close" or "loving," he interprets it as a desire for sex because of his conditioning.

Wendy, a thirty-year-old sex-worker who works in the field of "sensual massage," shared the following:

> *It was so interesting. Men would come in under the auspices of looking for a sexual experience, but what they really wanted was touch...the type of massage I practiced was set up as sex-work, but the sexual part happened only at the end of the session. Most of the time, the most important thing that was happening was the physical contact. I saw how hungry the men were for physical and nurturing contact. I often felt more maternal than like a sexual object, and more than any sexual act I would be a part of, their praise was often, "You have such a nurturing touch," or, "It feels so good to be touched."*

Having Mommy and Daddy in bed means that we are lying there not only with our parents, and each of their sexual histories, but with our parents' parents, not to mention our lover's parents. We know they're there with us because of the voices in our head that say, "You know you're too good for him." "Nice girls don't do *that*." "You know you have to make a man feel good to get love." "No sex before marriage." "Come on, just get it over with and he'll leave you alone." Whatever Mommy and Daddy felt about each other, and whatever they felt about us, is likely to be a big stain smeared across the sheets beneath us and our lover.

We are also in bed with a thousand years of conditioning about how sex is supposed to be—and the conditioning most of us received we could well do without! We are internally at the mercy of the viewpoints of our teachers and religious leaders. Who are our role models? Adam and Eve? Sylvester Stallone and Brooke Shields? Unless we had some exceptional conditioning, we're better off sticking with what our body is telling us about sex instead of what we learned, and thereby paving our own way.

Particularly if we were betrayed or sexually abused in childhood, suddenly we are in bed with Mommy or Daddy, Uncle Bob, the older sister we had a secret crush on, or the boyfriend who dumped us after the senior prom. We are ruthlessly denying them our love, adoration and affection because we are angry about things we may not even recall. Through our gestures we are unconsciously screaming, "Mommy, Daddy, Uncle Bob, etc.—I'm mad at you for not responding to me the way I needed you to, and this is how I will get you back." Our psyches cannot distinguish between the person next to us in bed and Mommy or Daddy, so we make our lover pay for it. We give our partner the message, "I don't trust that if I give my love to you that you will love me back. And if you betray me as well, I will be devastated."

We've got to realize that we are in bed with another adult—and yes they may be controlling, or passive, or angry, or timid *just like* our mother and father, but we need to accept that we *chose* this other adult! Bottom level psychological knowledge dictates that if we haven't worked out our stuff with mother and father (and that category includes almost everybody), than we're going to pick somebody to be with who is just like them. This person may look different from Mom or Dad, they may actually appear to be just the opposite, but when we get underneath the bedcovers with them, their similarity to our parents is often what we find.

If we are responsible, we will make this *our* problem and not our lover's. We may have found a mother/father substitute in our new partner, but that doesn't completely define who he or she is. We've got to give our partner the benefit of the doubt. He or she can't make up for everything we never got and if we try to make him or her do that (consciously or unconsciously), we will walk into a mine field.

One way to get Mommy and Daddy out of the bed is to pay enough attention to the kind of things *they* might say, as opposed to the kind of things *you* want to say, so that you can learn to distinguish their voices from yours. But don't give their voice too much influence over you. When thoughts, feelings and movements that you know are not your own start creeping in, you can go inside your mind and say to them, "Thanks but you can leave now. Sex is confusing enough without having *you* in bed too." You are fortunate if you can send Mom and Dad on an extended vacation so you can get down to having a honeymoon of your own.

Hint #4—Great-Grandma and Grandpa May Have Been Right

Great Grandma was looking over my shoulder like an angel when I first had sex, spreading her wisdom like a protective cloak over my naive body and mind. I didn't wait for marriage to have sex, but I'd like to think that I "lost my virginity" the right way. Fifteen years old and totally in love, my sixteen-year-old boyfriend and I planned our first sexual encounter well in advance. We talked about how scared we were and how we didn't know exactly what to do. I waited till my parents left town for the weekend. On the appointed night I made him dinner—

eggplant parmesan, garlic bread, salad and sparkling grape juice for a toast. We took a bath together, and then made love—clumsy love, terrifying love, but the sweetest kind of puppy love.

Actually, I never knew my Great-Grandma, but if I had the privilege of talking to her about sex, my guess is that she would probably say something like this:

- *Approach sex elegantly.* Waiting till you're sloppy drunk enough to have the courage to seduce somebody at the bar, and then having an orgy you will only half-remember the next day, is inelegant. Provoking a fight with your husband to get his attention, getting into a raging battle about nothing, just so you can make up and make love when it is over, is inelegant.
- *Only sleep with people you really care about.* And if you're married, don't sleep around. It is never worth it. If he finds out, he'll feel betrayed, and if he doesn't, you're stuck with living a lie.
- *Remember, a good man and a good woman are hard to find.* When you find the one you love, don't let them go easily. People tend to get upset by the small differences between them—thinking that no sex for a month means that the whole sexual relationship has gone to pieces, or that just because he gained ten pounds means he's no longer sexually appealing. Remember that you're both going to grow old, change and mature. Remember who you fell in love with in the first place.
- *Leave a little formality in the relationship.* At the turn of the century we were probably a little bit conservative, but men treated women like women, at least publicly, and women treated men like men. If you're married, repeat your wedding vows to yourself silently once a week. Recommit to the relationship. You're together to

love, honor and respect each other—consider what this means.

- *Remember, sex for men and sex for women are very different things.* Stay with it until both of you are satisfied. In the long run, you'll receive greater satisfaction when both you and your partner enjoy sex.
- *Be romantic.* Don't just spend Saturday night in front of the television set with a pint of ice cream as the high point of the night. Make dates—even though you live with him or her every night of the year. Take a shower, put on something nice. Hold her hand. Tell her you love her even if she already knows it. Open the car door for her (if she doesn't mind). Surprise him with a nice dinner. Go to the movie *he* wants this week. Pay attention to the details. There's more to this than meets the eye - it's a way of expressing your regard for each other.
- *Be spontaneous.* When you're my age, your fondest secret memories will be of the times when you took a risk and were spontaneous. What appeared to be major breaches in social protocol at the time come to mean nothing years later, except that you had enough sense to humor yourself and take a chance. Have a pajama party together and cuddle under the covers. Surprise him or her with a full body massage. Go out for a date having no idea where you're going when you get in the car. Have a romantic picnic in the middle of the city.
- *When things are bad and they have been for a long time, be honest about it.* Instead of being complacent, do something about it. You don't have to have an affair to shake up the marriage—just be straight and direct. (But remember, honesty doesn't mean making a billboard that lists every large and small fault of the other.) Or, see a counselor if necessary. If there's nothing left between the two of you besides codependency, reconsider the

relationship. When you're my age and you're looking death straight in the eye, you're going to want to know that you've used your years well.

- *Make love.* This requires giving your whole self, and if you don't give that much you're always going to sense that something is missing. Reveal yourself—dare to be foolish, shy, funny, exuberant, wild, talkative or silent.
- *Beauty is only skin deep.* Every man is going to develop a pot-belly and every woman's breasts are going to sag. If you're so vain as to reject somebody because their nose is too big or their breasts are too small, you're not mature enough to be having sex in the first place!

Everybody has a Great-Grandma or Great-Grandpa inside of them. These are our internal wisdom-figures who view things from a platform of a lifetime of experience. The reason we pretend we don't have their wisdom is because to own up to it implies a lot of responsibility on our part. It's much easier to stay ignorant and dumb. Take a look at your relationship from their perspective. Approach sex from there. It's guaranteed to change your attitude!

Hint #5—See Me, Feel Me, Touch Me, Heal Me

You can be in relationship for a long time together and still miss out on a lot of love that can be shared between you and your partner. To extract the full possibility of what relationship can offer, you have to be willing to *give yourself* to the relationship, and also to your lover's well being. Seeing, feeling, touching and healing your partner are good places to begin:

See me. I would always be so disturbed when a new boyfriend would say, "You're just like an angel. You seem so perfect to me." That was my red flag. If that was all he saw in the beginning, I knew that soon enough I would be the wicked witch of the West in his eyes.. and sure enough!

Take a good look at the one you're with—see them for themselves, irrespective of what they will or will not give to you, why they aren't good enough for you, or why they are too good for you. Make a consistent attempt to take note to yourself of your partner's fine points, but see the rest too. See the mole on her left cheek, see the part of him that is still five years old, see the extra five pounds on her thighs...and also see how brilliant she is with the kids; notice how well he does his job, or how he makes people laugh even in the most unlikely of circumstances. See her "Little Princess Game" or her "Save-Me" drama, see the remainder of his "Mommy-take-care-of-me-I-can't-pick-out-my-own-clothes" personality from childhood—and love him or her anyway. You're not going to find someone better, because a healthy sexuality has to do with *you*. You can trade in a clingy lover for a distant one, or a "little boy" husband for an aloof guy in a suit and Porsche, but when it comes down to it, you have to come to terms with your own sexuality.

Feel me. Pay attention to feeling. Take a step into your partner's shoes and try on the way he or she sees the world. What are the things that matter most to him or her? Why does she get so upset when you show up a half-hour late for dinner? Why does he have such a hard time *feeling* anything? What makes her cry? Get a feel for what moves his or her life, and don't judge it. Just come to know it.

Women, especially, want to know about feelings. "How do you feel about me *this week*?" she asks, while her husband is saying to himself that he has been telling her the same thing every week for the past twenty years. Tell her again.

Men can go either way. Some are relatively disinterested in feeling: "It's understood that we love each other—we don't need to talk about it all the time." "We're together, aren't we?" "Of course I love you." Other men are highly emotional and have a lot of feelings they love to share. Some don't know how to share feelings. Others don't reveal feelings because nobody has ever asked them before.

When it comes to feelings, there's got to be some give and take. It is reasonable for a man to not want to analyze the entire act of sexual intercourse five minutes after it is through. On the other hand, men *do* feel, and women know it. Women want to know what their man is feeling, and that's reasonable once in awhile.

Touch me. Bring touch into your lives together, and not just on a time schedule. Granted, there is nothing worse than a couple who constantly appear to be having sex in public, who can't go to the bathroom without the other, who can't drop each other's hand to have a conversation with somebody else...but for most people, a small affectionate gesture can really soften them and open them up to relationship.

Have sex with him or her, but also hold her hand, stroke his hair, touch her cheek. Touch communicates affection. It is easy to think, "She knows I care about her," but each time you cuddle up with her, you're expressing that care, and that is what brings so much aliveness to the relationship moment to moment. Pay attention to what he likes. If he curls up like a happy cat every time you rub his shoulders, do it for five minutes when he comes home from work. If you know it means a lot to her for you to rub her belly—rub it, and learn to like it simply because she does. Most importantly, do it without keeping an emotional bank account of accruing debts.

In sex, touch your lover how he or she wants to be touched. I can't tell you how many clients come in complaining about their sex life. "You know, my husband has never touched me the way I want him to. I can't believe he hasn't caught on after all these years?"

"Have you told him how you want him to touch you?" I ask.

"Well no, but I think he should have figured it out."

So they waste fifteen years of their sex life together because they're too scared to have a conversation about touch. Yes, maybe he has to swallow his pride to get the words out. Yes, maybe she'll be a bit embarrassed to accept that she hasn't been touching you the way you want her to all these years, but in the end, everybody will be better off.

On that note, take the time and the courage to experiment to find out what your partner likes—don't just sit there complaining that she is sexually frigid because she doesn't respond to your touch. Maybe that's because your particular way of touching her doesn't feel good to her. If you must point the finger of blame, point it at yourself and ask her what she wants—you will probably be surprised at the results.

Heal me. Make a commitment to help your partner heal. If you've done all of the suggestions above, you are already healing one another. Healing happens through the seeing, feeling and touching. Everybody has a deep yearning to be accepted *as they are.* They may never articulate it that way—they may think they just want to have a good screw. But when you're willing to make a commitment to heal, healing happens.

Be patient. Most men and women have had their sexuality wounded to some degree, and most need healing. People will start to let go and be more free in their sexuality when they really trust somebody else; and for some that will happen in two months, six months or a year, and for others it will take ten years. Be patient. When she's talking to you about what her father did to her as a child, or when he needs to tell you one more time how his mother smothered him and turned him into a surrogate husband, just listen. It will pass. It's a drag to have to be patient for his father's deed or her mother's attitude or the culture's prejudices or the religion's dogmas—nobody wants to pay those dues. But it's got to be done. If you care about your partner enough, it's the only way to go.

Heal your lover by being different than the others in his or her life who have hurt them.

Hint #6—The Body Knows

The body knows. The body knows pleasure—deep rapture that transcends the ecstasy of a ten-second orgasm or an hour-long massage. The body knows how to love; it knows when somebody needs a hug and when they need a little space. The body knows how to breathe deeply so that when you wake up in the morning you are rested and refreshed even if you had a late night. The body knows when you need to eat, when you need to sleep, when you need exercise. The body lets you know you're alive when the world is making you question that.

Your body knows everything you could possibly ever need to know about sex. The only reason that people need to read books about sex is because they are stuck in their heads. The modern world is a culture of walking heads, full of backwards ideas about sex. Heads create missiles, bodies create love. The body was born with knowledge of sex, and not only of copulation, but of bliss, rhapsody, freedom of movement.

Just as the mother, when left to her own wisdom, does not need a medical doctor to tell her what to do when her child comes out of the womb, or a book to teach her to breast-feed, the body contains the full wisdom of our sexuality. The exact knowledge of fertility, sexual cycles, breath, relaxation, pleasure—everything we need to know to take ourselves and our partners to the heights of ecstasy and far beyond—is fully encoded in every one of our cells. We can easily deduce that there was no sex education teacher telling Adam and Eve how to copulate. They knew. The body knows, animals know, flowers know.

In bed, the body knows how to give and how to receive. The body doesn't think, "If I touch him this way he might feel good," "I had better touch this breast as long as I touched the other one," "I think ten minutes of foreplay is enough," and so forth. That is how the mind thinks. The body thinks through instinct, and finding the mind of the body will unlock that. The body will guide you in sex far better than any plan or technique. It is through stripping ourselves of the straight-jackets of our conditioning that we get down to the bare essentials. We stand naked in our hearts and something real is exchanged. It is the body that will take us into our lover's heart. The body is not separate from us—it is who we are.

Every atom of the body is vibrant and alive. Most people have never even questioned how the inside of their bodies feel. When they think of the body they think of the skin. They don't even know that their heart lies in the middle of their chest, or where their liver is. They say they have a headache but if you ask them where it is they can't describe it.

The body in bed is always giving us signals (where to touch and how to touch our partner), but we have to be able to listen to our body to be able to detect these signals.

To access the knowledge of the body, we need to throw all of our previous ideas about what we know sex to be out the window, even if we've been sexing for the past forty years. When you forget everything you "know" about sex—how often you "deserve" to have it, why you should have it, or something somebody once told you about how a woman/man likes to be touched—then you make room for the body's signals to be heard.

Next, you start with the basics of "body listening." Try this: decide for a day that you will only eat when you are hungry, and pay attention to your body. Of course, the mind will tell you to eat breakfast at eight, and lunch at noon, but listen to the body. What does it say? Once you know what your body says in broad daylight, you can start to become more familiar with its subtle sexual signals.

Listening to your body in bed means that when your lover asks what you want him to do you don't say, "Oh, I'm sure whatever you do will feel wonderful." Bullshit. He or she asked you what you wanted, don't pass up the opportunity, and don't think of what felt good the last time and tell him to do that. Listen to your body—what feels, what moves, what calls out. Follow that, even if you don't understand its messages. There is a wisdom there that you may have never known about before. Find *that* and you will be happy. Find *that* and your inner world will open up. Find *that* and you will feel the inside of your body teeming with life and passion and lust and hunger.

When you're starting to be sexual, remind yourself, "*My* body knows," and keep deferring to it again and again. If body is saying "stroke his forehead," then stroke his forehead. If body says "open your eyes," open your eyes. It's not that you'll be absolutely clear on every message that your body is giving you, or that what your body knows will always be exactly in line with what your lover's body knows, but you will grow to learn how to read the body by making an agreement to listen to it no matter what, and then paying attention to the response.

Another hint is to pay attention to what your lover's body knows. Don't just listen to what he or she says, watch his or her body. Note the subtleties of its movements—the way the breathing changes, the posture of the mouth, the look in his or her eyes.

Real sex comes from the body, and only by getting *into* the body are you going to find it. You don't have to be perfect, and in fact you won't be. Just get out of your head and let the body show you a thing or two.

Hint #7—Real Men are Irresistible and Real Women Want Real Men

For starters, who are these "real men" who are so irresistible? Let's face it—real men are not Clint Eastwood and Robert Redford. The models we are given for real men are usually a bunch of oversized, impossibly handsome, patronizing dudes who treat women like they were paper dolls.

Real women want real men. And what's a real man? A real man is a man who is willing to reveal himself. You certainly don't have to be a SNAG (Sensitive New Age Guy) to be a real man—in fact, that can be a turn off. But real men are willing to show up as people with feelings, people with needs, people who can admit to insecurity, to needing their woman. A real man is willing to be present in bed with his woman; not to take over and show the woman what an amazing Casanova he is, but to be vulnerable, to exchange. Real men are strong but they can also be soft. Real men are secure in their appreciation of themselves. Unfortunately, a lot of perfectly fine guys believe that just because they're forty years old and have been through two marriages that their new girlfriend is going to leave them if they don't give her multiple orgasms the first night they sleep together. Real men are men who aren't afraid to be just as they are. That doesn't mean they show up without taking a shower and in their grungiest clothes just to show a woman they are not afraid of what she thinks of them. It does mean they can show up as an ordinary person and not as an act. A real man lets a woman know the things he thinks about, and dares to share a feeling (she'll love it). He can "get real" with her. Nothing is more annoying than talking with a man who refuses to let his guard down. There he is being Joe Cowboy or Mr. Stud, or whomever he imagines will impress you, and you just want to shake him and say, "*Who are you* beneath all that?"

Men are always putting on all kinds of acts to please and impress women. Unfortunately, who they're really trying to impress is either Mommy or the big Mama in the sky that's supposed to take care of them. Society has conveyed to them that if they're aloof, if they're in control, if they're on top, if they're arrogant jerks who don't need anybody, that this will make them attractive. Perhaps if a man wanted to date a female character on the soap opera *General Hospital* these attitudes would make him attractive to her. But we're talking about *real women*—and what real women want is somebody to *be* with. Somebody they can sleep with *and* talk with, and have fun with, and cry with. Yes, it's a tall order, but not really. Over and over again, you hear women saying, "I just want *him*."

One day, on the way back from picking a friend up from the airport, she was talking to me about her long-term relationship. She said, "I told him I was at the end of my rope—that after all these years all I really want is *him*." I chuckled, and when she appeared offended as though I was laughing at her, I quickly explained that I was laughing because probably 90 percent of the women across the world say the same thing! I wanted to know more. "Just what do you want?" I asked her. "More time? More conversation?"

"It's nothing measurable," she responded, "it's just that he holds back—he has never given *himself* to the relationship, and actually that's all I want from him." And that's not easy for most men. They themselves barely know what it is she wants, and they have never considered sharing it with anybody, much less a woman.

Well guys, this is what your woman wants, and if you give it to her you're going to find one happy, radiant woman who will be only too glad to put up with your idiosyncrasies because what she is getting is *you,* and she is well aware that she is holding the prize that most women want. You'll be able to get away with all kinds of things when you give her the one thing she wants. Imagine—all you have to do to please your woman is be yourself. Now it's up to you to figure out how to do that!

Hint #8: Women: Get Out of Your Head

> *...Oh gently, gently,*
> *let him see you performing, with love,*
> *some confident daily task,—*
> *lead him out close to the garden,*
> *give him what outweighs*
> *the heaviest night...*[5] - Rainer Maria Rilke

No matter how extraordinary a man you're with—how enticing his touch, how much he cares about *you*, if you're stuck in your head, he could be breathe adoration into your ear and you wouldn't even notice.

Women need to get out of their heads. They are sitting under ten thousand years of conditioning that tells them that their worth is equivalent to their breast size, and that they are valuable human beings to the extent that their husbands are sexually satisfied. Women's heads are as full of ideas as men's are about how they're supposed to be in sex, how they're supposed to touch their man, and how they're supposed to feel, that there's no space for *her* amidst all the wreckage.

Women's minds tend to think in more or less the same way. Women's minds say things like, "All men are like little boys." "He's just like the rest of them." "Why doesn't he share his feelings?" "If he really knew what I wanted from him he'd be out of here in a minute." "He doesn't understand women." "If my thighs are too fat he's not going to be attracted to me." "If only I had so-and-so's legs, and so-and-so's chest, and so-and-so's eyes...then things would be different."

And this confusion and self-deprecation is not all that's in the way. Demi Moore and Arnold Schwarzenegger are in the way. Books that tell her how to increase (or flatten) her breast size in thirty days are in the way. *The Enquirer* is in the way. Nutrasweet™ is in the way. Cellulite treatments are in the way. Child pornography is in the way. Internet sex is in the way. Advertisements that use larger-than-life-size men and anorexic women to sell phallic-shaped cars are in the way. Miss Manners is in the way. The voice in the head of one woman I interviewed who said, "If my husband knew I had a fat ass he wouldn't love me" is in the way. My former roommate—six-foot-tall, one hundred and twenty-five pounds—who would look in the mirror and say, "I'm getting fat," *is in the way* of a woman's sane relationship to her sexuality.

So many women are simply having sex in their heads, their bodies following along by command. The importance of getting "into" your body is discussed throughout the book because people—especially women—need to experience this. The way you "get it" is by doing things with your body besides putting fat-free ice cubes into it and calling them dessert. Consider an option to spending $80 a month at the beauty parlor just so your hair can look like a frosted flake. Get your hands dirty—dig in the garden. Get your hips swinging—take belly dancing or Flamenco. Get some clay or paints or whatever you have never told anybody you really want to do and start shaping it. Don't just watch *Thelma and Louise* on the movie screen—take your partner and go for a joy ride, or kidnap your man from the office on Friday with the car packed with sleeping bags, Brie and Cabernet, and go for it! (Just don't drive off the cliff.)

Once you've done all that—which is not easy considering that most women are so weighted down with expectations of themselves that they seem to be dragging around a 200-pound boulder attached to their ankles—then the challenge that awaits you is in the bedroom.

You've got to show him who you are, and don't pretend you don't know what I'm talking about. You may have never shown it to anybody—not even the mirror when you're home alone in the middle of the afternoon on a weekday—but you know what's underneath there. You know that you're far more than the passive recipient of your husband or boyfriend's advances. Dare to let your body guide you in sex. The first five or 105 times, your mind is going to be screaming at you: "Who do you think you are?" "He's never going to go for this!" "You're faking it—you're not really this free." "Good women don't do these things." "I just *can't do this.*"

Do it anyway. And be warned, if you've been following the same sexual routine for eight or twenty or forty-five years, he *will* be surprised, and perhaps intimidated at first. If he makes fun of you, stuff a pillow in his face and stay in touch with what your body wants. *Your body* knows enough for the two of you. You've both been trained that he's supposed to know what he's doing, but really the secrets lie in *your* body. Yours is the body that knows how to give birth, that sheds an egg with each new cycle of the moon; your body that has more than enough knowledge to tap into the highest states of ecstasy. You just have to figure out the way to get to it, and the way you do it is by saying "yes" to your body and "no" to your screaming mind, thereby paving the way for both of you.

If your mind just won't let you do it, embark on a discipline in which you cease to stop criticizing every wrinkle, every inch of cellulite, every hair that grows somewhere you think it shouldn't. Let your new lover see you without your makeup and realize he loves you anyway. Refuse to torture, contort and manipulate your body to make it how you think *he* wants it.

And by the way, don't worry what size your body is. Haven't you seen movies of large, voluptuous women who really know how to *dance?* Can you honestly say that they don't have sexy bodies? They put the white-bread pixie-stick bodies to shame! Your problem with your body is in your head, and although it's a tough one to get over—*know* that it is in your head. (And if it's coming from your lover, it's in his head too.) When you both get out of your heads and down into your bodies, you're going to forget every criticism of your physique that has ever crossed your mind.

Hint #9—Men: Place Your Attention on Your Woman

You'd be amazed at what happens to a woman when you place your attention on her. A man's attention and adoration can make an ordinary woman into a raving beauty. Surely you've had the experience of meeting a woman and finding her only moderately attractive, and then getting to know her, or falling in love with her, or watching her with her children and suddenly finding her exquisitely beautiful—that's what attention does to a woman. It's like somebody turned the evening light in her direction—suddenly everything glows with a fading dark beauty.

On the flip side, ignoring your woman, or putting your attention on other women, turns an ordinarily attractive and charming woman into a madwoman—the consuming witch you've always dreaded and probably experienced. Which would you prefer?

A lot of men put their attention on the Wall Street Journal, or on the football game, or on their six pack, while a gorgeous and sensual woman is sitting there drying up beside them. Attention is food for a woman. Putting your attention on your woman means paying more attention to her than you do to the television set when you get home from work. It means noticing what you're eating when you know she's spent the afternoon cooking it and is anxiously awaiting your response. It means that when you give her a kiss before you go out the door that you are *in* the kiss and not already in the car. It means complimenting her when she looks nice instead of assuming that she already knows it. Attention makes a woman ripe and juicy, so when a man complains that his woman is dull, I ask him how he treats her.

Putting your attention on your woman in bed can bring a revolution to your sex life. Let's face it, sex in the modern world, and almost anywhere in the world save perhaps ancient Egypt, is centered around a man's orgasm. Most people do not think that sex has "happened" unless the man has ejaculated. On the other hand, a few good groans from a woman—faked or not—are often considered sufficient evidence that her needs have been met.

To put your attention on your woman in bed means to consider centering your lovemaking around *her* experience. No, you won't get ripped off, and no, you don't have to give up your own experience to simultaneously allow yourself to tap into her experience. But women hold the key to lovemaking for both man and woman, and the way a man turns that key is by placing his attention on the woman. It may take practice, and a willingness to risk control of the situation, but it's a guaranteed formula. When your woman trusts you, and can get out of her head enough to get into her body, you both go for the ride. The woman's body holds many secrets that can bring both she and her man to heights of lovemaking in a way that "getting your rocks off plus a few extra kisses" never will. Unlocking a woman's passion has the capacity to take you both to places you've never been. Men who yearn to understand women have often asked, "What's going on when a woman is having an orgasm? I want some of that!" And he *can* have it, but he has to be willing to pay the dues of giving her the attention and patience she needs to reveal herself.

To place your attention on your woman during sex you may try any or all of the following:

- *Let the woman guide you*—in every way. You allow her to guide you when you are willing to follow. Watch her eyes, watch her body, watch every movement, everything.

In doing so, you will intuitively know how to touch, where to touch, when to touch her.

- *Have sex on her timing.* It is her body that is wise beyond words in its ability to conceive, to carry a child, to make its food. Imagine what else it knows.
- *Talk to her.* Don't assume you know what's going on for her.
- *Give up the goals.* Lee Lozowick, author of *The Alchemy of Love and Sex,* has said, "A good orgasm is never the end of the road, although sometimes it's a detour." There's nothing wrong with orgasmic sex, but there's far more to sex than just orgasm. If you can shift your attention from your goals to your woman, you'll discover just that.
- *Keep your eyes open and the lights on,* from time to time.

A middle-aged truck driver who spoke with me about this said, "When my attention is on me, it's fucking. When my attention is on the woman, it's lovemaking."

And watch out men, because if you really do this, who you're going to get is a real Woman. You can say good-bye to your Barbie doll image because who you're going to meet in the flesh will blow you're mind, if you're lucky. And if you're afraid, you should be, but keep your attention fixed on her anyway. You'll get over the fear, and after however many years you've been trying to get it right, you'll enter the domain of something new in your sex life.

Men often complain about how much attention a woman wants, but they neglect to consider what she becomes when she gets that. A man's attention turns an ordinary housewife into a devoted lover, a consort, a true sweetheart. There is a real woman in every female person walking this earth, only most men have never discovered her. Do you want to die without having dug for gold in that mine?

Hint #10—You Can Have a Passionate Life Even Without a Mate

> *You must be a human being before you are a man or a woman. You must be a Lover before you are a sex object. You must know joy before passion, generosity before greed, gentleness and compassion before superstition and prejudice.*[6] - Lee Lozowick

A bona fide California new-ager once proudly boasted to me that her sexual preference was "*universeogamous.*"

"What's that?" I asked her, choking on my judgments.

"I don't limit myself to people," she responded, "I can have sex with trees, stars, the wind, animals, the Goddess..."

"Give me a break," was my response to her at the time. But, thinking it over more seriously, I realized that she had a point. We don't have to wait until we are in a sexual relationship to lead a fervent, enthusiastic life of sensuality.

We can have a sensual relationship to many things (dance, food, friends, nature), or we can have a more neutral relationship to them. Our relationships with our friends, our caring for our children, the way we dance, our ability to enjoy food, music, art and nature are all functions of sensuality *and* sexuality. Sexuality must be seen as more than copulation or we will miss the point.

There are sensual, passionate and brilliantly creative individuals who are celibate for years at a time—either by choice or by default. Being celibate does not mean being dead. Believe me, your problems don't suddenly go away just because you're sleeping with somebody. Not only can you learn to live with celibacy, but you can learn to be happy in your celibacy. And no, the answer does not lie in masturbation several times a day, but rather by developing a passionate relationship with life. This is done by listening to your body deeply and responding to its messages, through cherishing the close relationships in your life, and by living fully and not settling for mediocrity. Sexuality lies in your body, and even the sweetest of lovers do not have a real sex life if they're not in their bodies.

If you want to lead a thoroughly sensual life, don't wait for Mr. or Miss Right to come along. There are many practical means by which to develop intimacy in your life besides coupling. Observing children is an excellent way to get in touch with your sexuality. Children are naturally sensual—they are not ashamed of their bodies. They are present, in the body, and full-on in whatever they do. Give yourself the gift of spending quality time with children. Many people claim they have no touch in their lives, but their nieces and nephews live five miles down the road and they don't spend any time with them. If you don't have your own children, you probably have a friend who would be only too glad to have you do some free baby-sitting. And don't just watch over children, bond with them, get in the sandbox with them, stick your hands in cookie dough together. Don't grasp at their affection, just be with them in their world and you will effortlessly find your way into your body.

Creativity is another excellent channel for sexual energy. All that potential sexual energy makes for excellent painting, singing, music-making, writing—whatever it is you've always wanted to do but never had the guts to try.

You may also discover that a great deal of satisfying intimacy is possible with peers of your own gender, the subject of *Hint #11.*

Many people who are not in a relationship are terrified of intimacy—period. There is nothing wrong with being afraid of intimacy, whether you are aware of it or not (many people who are afraid of intimacy tend to simply think of themselves as "independent," or "loners" by nature), but if you do hope to find real relationship in your life, you had best begin by learning to relate to the people in your immediate environment. Develop deeper bonds with the people you are close to. Dare to put your arm around your best friend. Ask for a hug when you need it. Take one small step past what you are accustomed to—no harm will come of it. If you cultivate sexuality and intimacy within yourself and with those closest to you, you will be ripe and ready when and if an intimate partner comes your way.

Hint #11 - Cultivate Same-Gender Friendships

> *An objective examination of most of the world's cultures will reveal a distinct division within the culture. There are actually two cultures coexisting within the society: one for the men, and one for the women. This distinction is most obvious in the division of labor, but it runs much deeper than that. There is an inherent difference between the spirit of man and the spirit of woman that demands respect.*[7]
>
> \- Lee Lozowick

I never knew the pleasure of spending more time with women than with men until I started traveling and living in places where men and women don't generally associate socially as friends. Until that point, I usually had one or two good female friends and a slew of male friends. My male friends would always validate me, make me feel beautiful and let me get away with murder just because I was a woman. Male friends didn't talk back to me. Nice as this was, I didn't know what I was missing out on.

When I went to India, all the women would hang out together. From a man's perspective, the women were locked away in the house, but when the men left, it was hardly a prison! The women would often laugh until they cried as they cooked a meal together. They would go over to one another's houses to do the laundry just for the company. One day when I was attempting to dress in an Indian sarree, I had wrapped myself in a knot. Shyly, I knocked on the neighbors' door. Though we had never spoken before and none of them spoke English, I conveyed to the young women, through hand motions, that I didn't know how to wrap all that material around me. Within minutes, several lovely women were in my room, wrapping the sarree around me, then combing my hair, putting oil on me...it was delightful! This is how women are together quite naturally.

Then there were the Indian men! It was the first place in the world where I had seen heterosexual men regularly walking down the street hand in hand, or arm in arm. I must say it was *impressive* to see men able to love each other and show it—an unordinary sign of real masculinity.

Now obviously we don't live in India, but we stand to gain a great deal by paying a lot more attention to relationships with members of our own gender. Most cultures around the world recognize the fact that women and men orbit in different universes. Whereas sex and marriage usually happen across genders, a lot of social time is spent with members of one's own sex. Only men can really teach a man how to become a man, and the same is true for women. Forming close relationships with people of your own gender gives you a tacit sense of strength and steadiness that you just don't get if you limit yourself to cross-gender friendships.

Generally speaking, forming same-gender friendships is easier for women. For men—particularly men who want contact with other men—the only socially approved ways to make physical contact are by getting so drunk that they end up putting their arms around each other, by butting into each other in football games, by giving each other high-fives, or by wrestling with one other on the floor. Women are more in the habit of dressing one another, hugging each other, and giving each other friendly pats on the back. But, as a whole, many men and women are each afraid of their own kind.

Trends in the contemporary New Age scene that aim to break the same-gender taboos, haven't quite hit the mark either. There are men's groups that meet out in the forest where the guys howl like cavemen and try to get themselves to cry; while women are going around hand-in-hand in Goddess garb calling themselves the Divine Mother. Sticking to your own kind implies something else, however. For men it means choosing to spend the afternoon with your son and his friends, or with your old buddy, and going fishing, or to the movies, or camping for the weekend...just because. (There is a fair amount of contact and intimacy that can happen in an afternoon of Frisbee.) It means valuing the role that other men play in your life.

For women it means making plans to spend more time with your women friends—especially if most of your acquaintances and friends happen to be men. Plan a women's night out, or "in." Wash each other's hair like you did when you were teenagers. Exchange back massages. Have a slumber party. Cook a meal together and eat with your hands.

Women walk around complaining that men can't understand how they feel (which they can't), when all the while they could be talking to other women who know *exactly* how they feel. The same is true for men.

A friend of mine recently called me up, upset about the pending break-up of his long-term relationship with his partner. When I asked him what the central issue was, he told me that she felt that he was not sensitive enough to her emotional life. This surprised me, as this particular friend is one of the most emotionally sensitive men I know, so I asked for details. Apparently, there were aspects of longing, depth, sensitivity, and pain in her that, although he was able to understand in principle, he was unable to feel in the *way* that she felt them, and she felt abandoned and isolated in the relationship because of it. Hearing this, a wave of sadness and frustration washed over me, both for the feelings of inadequacy my friend must have felt, as well as for this woman's lack of understanding that what she wanted from him she was never going to get from him or any man (ever) since the way that a woman experiences the deep feminine in her body will not and cannot be felt in the same way by a man. Similarly, women will never know the bodily experience of a man's sexual drive, or his orgasm, or his need to be alone with his feelings.

Women should talk with other women about things like mothering, sexuality, menstruation, health, and how to live with an emotionally distant man. A man needs to speak to other men when his wife's post-natal sex drive disappears, when there are difficulties at work, or to discuss the challenge of expressing affection to his children. If he feels smothered by his female partner he should seek advice from his best male friends or wiser, older men, instead of ignoring her or running away from her. If adults don't talk to members of their own gender, women will try to teach men how to handle things a woman's way, and men will try to tell women how to be men.

When problems arise between men and women, instead of fighting it out with their mate, people in many cultures will instead turn to their own kind for help. The following example is from the Pygmy tribe in Uganda:

> *What usually happens when a husband and wife fight...is that they are encouraged. If a man is about to hit his wife, the other [men] will give him a stick and say, "Hit her with this. You are a strong man. You can kill her!" By this time, the husband already feels a little ashamed. The others group around and call out, "O.K.! Go! Go for it!" And then he realizes how foolish he looks. They end up making a joke out of it, a sort of soap opera. Then everybody claps, and they are happy.*[8]

Obviously, this will show up somewhat differently in contemporary culture, though I recently heard a comparable story from a co-worker. Apparently, she and her husband were fighting about some sexual problems, and although there was no physical violence, they were obviously stuck in the rounds and only wounding one another further with their insults. Suddenly, there was a knock on the door, and there stood my co-worker's sister (who lived in the apartment upstairs)—a mature, sensitive and ordinarily soft-spoken woman—with two knives in her hands. She said to them, "If you really want to kill each other, use these! And if you're not dead afterward, Jack, you go talk to your friend Andy, and Elisa, we'll work this out between *us*—the problem here is not each other." They both agreed later that it was far more effective to talk some of these things over with members of their own gender than it was to duke it out between themselves.

Cultivating same-gender friendships means showing your face to your peers who see through your tricks. Men and women let each other get away with a lot of stuff that friends of their own gender would never let pass. The question is, however, do we really want to be real? If the answer is yes, then we have to risk hanging out with those who are going to demand this of us.

Hint #12—Get on Your Knees and Worship Your Lover

> *since feeling is first*
> *who pays attention*
> *to the syntax of things*
> *will never wholly kiss you;*[9]
>
> -E.E.Cummings

I dare you! Get on your knees and kiss his feet. Or serenade her beneath her bedroom window at midnight. You probably cannot imagine something more humbling. In many spiritual cultures, however, it is an outright practice to view your lover as God (or the Goddess). It's not that you pretend that he or she doesn't have faults and imperfections—you know that all too well. On the other hand, what you tend to forget is that contained within your lover are the seeds of the perfect man or perfect woman, and if that is what you make a conscious decision to relate to, that's what you're going to get.

Worshipping does not mean that you fill your house with candles and that you lay garlands of roses in front of your wife before she goes to sleep. Nor does it mean that you are expected to wash his feet! That could drive him crazy. But you can treat him or her as though he or she were the most precious thing you had ever seen.

Even if you would never consider getting on your knees, what I am suggesting is to experiment—even for one evening—with treating your lover like the God (whatever you conceive God to be) you worship...pretending like God himself or herself is with you in the form of your lover. You don't even have to tell him or her you're doing it, and it needn't be obvious. Instead, show up on time and consider doing what he or she wants to do. Pay attention to the details—how you prepare the food, or how you dress. Listen to what he or she has to say, and don't dominate the conversation or inwardly complain to yourself about how dull he or she is. If you go out together, don't look at other men and women if that's not what you would do if you were dining at a Chinese restaurant with God.

When you get into bed, touch your lover like you were touching God. If you got into bed with God, you probably wouldn't be thinking about whether God was going to take you to orgasm or not. It is unlikely that you would be thinking that God "just never gets it right," or "how come God seems to be having more fun" than you are? Your attention would probably be on giving as fully and completely as possible. Your touch would be pristine. Your mind would be on the other.

Similarly, receive your lover's gestures as though you were feeling the touch of God. After all, what does God really need from you anyway except for you to receive the love that is being offered to you? Allow yourself to be loved. There is no lack in the amount of love that is available; instead, psychological blocks often prevent people from receiving the love that others really want to give to them.

Pretending that your lover is God may sound absurd, but try it. Do it for a month, fully, with your heart and guts spilling out of your body. I guarantee you'll recommend it to others. Living in a culture where God is nearly obsolete (leaving only His or Her skeleton in the form of contemporary religion), worshipping anything is hard for most people to imagine, much less the one you love.

Worshipping your lover is an attitude of reverence. It shows up in small ways like laying out her towel for her when she gets out of the shower, or turning down the sheets on her side of the bed at night, or putting your arms around her when she least expects it. It may look like making him a cup of hot tea when you know he's stressed out with a deadline, or giving him a kiss hello that says to him that you really are glad he's home. Worship, in the form of service, comes from fullness as opposed to obligation. You make him a nice dinner not because that's what your mother always did and you don't know anything different, but because as an adult you see that is what pleases him when he comes home from work. Worshipping your beloved means approaching him or her in humbleness, recognizing that whatever faults he or she may have, you've got just as many.

I know a woman who used to pray to God that she would find fault only in herself and never in her lover. It is amazing how a difficult relationship can get totally turned around when you approach the other with this non-judgmental attitude. You are focused on your ability to love, instead of his or her inability to love, and as your ability to love strengthens, your lover responds in kind.

If you can stick with this approach to your lover, though such actions might feel silly to you at times, it is *you* who will ultimately benefit. When you treat somebody else with that much respect and reverence, *you* become all the bigger for it, whether he or she ever responds to it or not. It is *you* who will experience love pouring through you, replacing feelings of mistrust, resentment and criticism. The option to this is treating your lover with resentment, pity and condescension. Which would you prefer?

Most people can't imagine placing themselves so low and their beloved so high, but the few who have tried it have a love life, not to mention a sex life, that transcends all feelings of self-importance and self love. Taking on your lover as your Beloved is bordering on the mystical, but it is also very practical. You can know in advance that you probably won't succeed. Most of us are too afraid of the kind of intimacy and closeness that it would bring. But it doesn't altogether matter if you succeed. Simply having the intention to treat him or her as such, and taking any steps in that direction no matter how small, will make you a better person and give you a better love life. Your attempts will fall like a soft blessing on the object of your love.

The main connection between worshipping your lover and touch (aside from the fact that your touch will be far more sexy, nurturing, erotic and compassionate) is that when you treat your partner in this way, you *are* touch. How you are relating to him or her is an expression of what it really means to touch. The final chapter of this book is about *the way of touch*—touch as a mood and way of being which you bring to others. This "way" shows up through your gestures, but it is really who you are. Become that, and watch your problems dissipate into meaningless blips on the screen.

CHAPTER 7

Touch as Context

your slightest look easily will unclose me
though i have closed myself as fingers,
you open always petal by petal myself as Spring opens
(touching skillfully, mysteriously)her first rose[1]

–E.E. Cummings

Unable to find any cultural model in the Western world of a people who were not touch-starved, I began to watch those individual around me who were most "in touch." One man in particular caught my attention. He seemed to draw others to his company simply by his presence, and would leave them feeling touched in an inimitable way. To my surprise, closer observation revealed that he was rarely physically demonstrative of his affection. When questioned about this, he said that he had not been raised in a climate of consistent touch, and simply wasn't accustomed to it. How then, did he touch others so? His touch came through his eyes, that were ready to meet yours whenever you dared to look up and allow yourself to be seen. He touched by the quality of his listening when you spoke to him; his perfect attentiveness. He touched you in that whenever you needed help—even small help—he would suddenly appear to open the door for you, or he would show up with the last needed ingredients for a meal. He touched with his words, reflecting your point when needed, and refraining from speaking when nothing was required, never drawing unnecessary attention to himself. He touched with his smile, because you knew he meant it when he showed it, and also with his disapproval, for this was a sure sign that you had overlooked something important that was making *your* life more difficult. He touched you by the way he was with children, treating them with full interest and respect—not just occasionally, but always. He touched you because you saw he knew how to take care of himself, not demanding this from others. He touched you because his presence was touch itself. He didn't *do* touch, he *was* touch, and because of that, you felt touch in yourself.

Men and women such as these are the saints of our time, whether recognized by millions or known only within their own families. Though some give touch freely, such as the Hindu saint Mata Amritanandamayi who takes each of her thousands of visitors into her arms, holding every one of them and pouring her love into each, others, like Thomas Merton, who lived for thirty years hidden away in a cloistered monastery, similarly touch millions without ever seeing them or laying a hand on them. It is the context of a life lived in touch that affects those around them so strongly.

Joseph Chilton Pearce explains that people tend to be contextually determined, the context from which they are operating determines their response to any given situation. Of this he says, "Our culture breeds us to react to our outside context and so when we are in misery and pain and unhappiness, we try to change the outside context because we think we are determined by that context."[2] Although we insist on holding out for happiness until we have found the right mate, the right job, the right home, real touch comes from inside ourselves. Real touch is based upon the recognition that what we are ultimately looking for as adults will never, can never, be found outside of ourselves, that the *only* possibility for lasting and abiding fulfillment lies within. It involves taking the elements that are already here, and using our conscious capacity as human beings to masterfully direct them in such a way that love is produced.

Real touch is nothing more—and nothing less—than an opening of the heart and a relaxation of the mechanisms of ego that keep us from loving and being loved. It is a willingness to allow the body to open, both on the surface and deep within the cells, and to partake of a source of love that is continually available—not only from one's mates, children, and loved ones, but from what could be called God, the Self, the Great Spirit, Jesus Christ, the Infinite, or any name we choose to call the intangible reality that is the source of our existence.

The context of touch, therefore, is a way of life that rests on the premise of the inherent goodness present in each individual, as well as upon the recognition of an abundant presence of love. It does not naively propose that with a single kiss toads will become princes, but it does suggest that if we continually strive to understand the underlying suffering of the individual that is at the basis of his faults, and are willing to extend our love to him in spite of these, that the context of touch will be present, and healing will occur. To live life in the context of touch is to live in continual contact and intimacy with others—not exclusively in physical contact. Rather, the whole of one's being stays open to constantly being touched by and touching everything in its environment.

Whereas the context is extraordinary, a life lived in touch is ordinary. The context of touch is a mood of openness and a desire to bring love to the other, hour by hour, minute by minute. This vulnerability brings us immediately into the present. As we allow ourselves to be touched, we are suddenly met with reality unfolding in front of our eyes. We find ourselves without buffers, face-to-face with our child, spouse or friend. When we allow ourselves to be touched in this way, we realize that there is only the present moment—that our only "power" in terms of our capacity to make an impact on the world lies in the situation directly before us. We are either in touch with the present moment, or we are not. Once the domain of the heart is open, we don't need to be told how to touch, where to touch or when. Others don't need to convince us of the importance of touch. We know to touch, and we know that to live in a way that our touch is felt by others is the only way to really live.

True touch is not something that can be acquired, as much as it is found by relaxing into it. It is more passive than it is active. The touch itself is an action, but what gives rise to the mood of touch is a letting go of tension, and an allowing of ease, softening, intimacy, elegance. For those who have never let go, however, this is not always easeful. Some of us have hung on so tightly with all of our might that we have no idea what it means to let go—to just let go.

If only the individual could know that the pain he or she spends a lifetime protecting themselves from cannot compare to the suffering that the heart and soul already endure under the strain of their unwillingness to feel. If only the fire of the heart would burn strongly enough to melt the chains of fear that enclose it so that the individual would know, beyond the shadow of a doubt, that the greatest source of human suffering is our resistance to feeling the pain of our own hearts and of humanity at large.

Some would call it *grace*—that force which gives the individual the desire, strength and courage to move through resistance, see falseness with clarity, and feel the underlying pain that one must feel in order to become truly human. Others refer to it as wisdom or an inner knowing. Regardless of what it is called or where it comes from, this force or power is what we all seek; and although it comes from within, it is most easily accessed by touching others. When we touch others (and I don't mean giving them a polite and unfeeling kiss on the cheek), when we genuinely extend ourselves to them, we feel our own love, the love that is ever-present within ourselves, but is accessed through our willingness to give of ourselves.

Still, a life lived from the context of touch does not happen automatically. Rather, it is an ongoing affair. We move in and out of touch, get tastes of it and then fall back into habit or fear. Finally we find ourselves back where we started from, perhaps able to touch with a bit more ease.

Poets through the ages have described the hundred flavors of love. Just as the Eskimos have over a hundred names for snow because it is so essential to their survival (each name elucidates a specific quality, texture, size, degree of moisture, quantity, etc.), and as the Hindus describe one God by thousands of names, so when one lives in the context of touch, in communion with his or her circumstance, love becomes not one thing but many. Limited by their own minds and hearts, most people confine love to the romantic type, or the motherly expression, or perhaps to loyalty or commitment. But love can have qualities of fierceness, sorrow, compassion, wildness, rage, longing. A life lived in the context of touch is a passionate life full of many shades of love. This does not mean that the person experiences a continual intensity of emotional highs and lows, or always eats exotic foods or dresses in extravagant clothing; instead, his or her ordinary experience is alive with richness and depth, even amidst what appears to be a mundane life.

When a person shifts to the context of touch, a profound bonding is reestablished within himself and he begins to take directions from within, as opposed to relying solely upon his external circumstances. This in no way implies withdrawal from the world on any level, as his internal life remains in steady relationship to the world around him, but it does suggest that the source of his actions arise from an attunement to his surroundings, instead of an automatic reaction to them. When touch is the context of one's life, the individual's interactions with people attract love and brightness into his life. It is a magnetic quality that is capable of transforming the environment in which the context of touch is present.

The fact that thousands of spiritual, environmental and self-help movements are springing up across the globe, points to the first step in a shift of context. It is true that many people are participating in these groups for reasons other than they might believe (i.e., people say, "I want God," but the unconscious motivation is often, "Bliss me out so I can get away from this pain!"), and it is also true that many of these groups are fooling themselves about the lasting transformative effects of their work, but the very existence of these groups is a statement of the growing awareness of the undeniable and unavoidable need for change.

The shift to a context of touch is the change that people are seeking. Everything else follows naturally from there.

Therefore, we initiate that shift by touching one another in loving ways, and refraining from touching each other in unloving ways. We pay attention to our internal experience, accept our own bodies and our own humanity, and in doing so begin to understand touch. We come to appreciate that touch, or the lack of touch, is a constant in our lives, and that we are either taking responsibility for participating in it consciously or we are not. We touch and are touched and begin to perceive what the way of touch is. As we become more loving, others are naturally touched by the way we interact with them. People begin to enjoy our company without knowing why. We take interest in others, and we look for opportunities to serve and to touch them. To touch is to live, and through our touch we come alive.

THE ILLUSION OF SEPARATION AND THE REALITY OF INTERCONNECTEDNESS

> *A human being is a part of the whole that we call the universe, a part limited in time and space. He experiences himself, his thoughts and feelings, as something separated from the rest—a kind of optical illusion of his consciousness. This illusion is a prison for us, restricting us to our personal desires and to affection for only the few people nearest to us. Our task must be to free ourselves from this prison by widening our circle of compassion to embrace all living beings and all of nature.*
>
> -Albert Einstein

Most people believe that the human being is somehow other than, or separate from, what can be called God, Spirit, the creative force, etc. Just because somebody can say, "God is everywhere and in everything," or wears a fancy philosophy on her sleeve, does not necessarily free her from the deeply ingrained belief system that God is somewhere in the sky, and that human beings are down below, having somehow fallen from grace.

Notions of God as savior, of God as other, have been passed down from generation to generation, reflecting our alienation from truth, mirroring the mind-body split, and often being used as an excuse for the need to take responsibility for our lives (i.e., "If it's all 'God's will,' then I don't have to do anything.)" Whether spoken or unspoken, this separation from God is the underlying context of the human condition, its roots more ancient even than the failure to have bonded. When we feel ourselves separate, it is as if we have not bonded with our very source. This illusion of separation is the wound we all share. Even those individuals who had highly-conscious parents, and for whom the bonding process was never disrupted, also remain at the effect of this.

Believing we are somehow separate from or other than God, carries with it the implication that there is a force that is working against us, waiting to destroy us by death, and that we must therefore take care only of ourselves—defending ourselves, seeking to benefit ourselves, acquiring for ourselves everything we can get. If we are separate, then we have the "right" to destroy the environment, to do what it takes to make it to the top, to support a "survival of the fittest" mentality, to dump a partner who doesn't suit us, to sleep around, to work for companies that exploit their workers, and to take what we can get, regardless of the implications of our actions on others. If we believe we are separate from others, we do not go out of our way to touch others, as we are too busy protecting our own self interests.

This instinct for survival, based upon every individual's belief (including much Darwinian thought) that he or she is separate from every other individual, underlies all forms of abuse, aggression and self-protection leading to isolation, loneliness, depression, withholding of touch and the other shared ailments of our time.

Concerning the prevalence of abuse and violence in our world, if we truly understood that we were not separate (that who we essentially *are* cannot be destroyed), there would be nothing whatsoever to defend, and all tendencies toward abuse would cease. But we don't understand this. Instead, we live our lives based on this false assumption of separation, often dying without ever recognizing or questioning its validity.

In this vacuum, people attempt to harness God's help by turning to popular religion. They want something, anything, to make them feel cared for, to make them feel loved and secure. Oftentimes, however, what they are actually following is man's mistaken interpretation of religion, instead of God or the Universal Law, and thus they fall into the trap of dogma and ritual while missing the point of *experience* entirely. Due to confusion, ignorance and the misunderstanding of "tradition," they turn to a false (or dead) God, never finding the salvation they seek as they are looking in the wrong place. When God becomes a savior who is separate from us, instead of a living presence in the nooks and crannies of daily life, it is an indication that people have substituted the necessity of intimacy for an idea of salvation.

God does not live in the sky. If we believe that He, She or It exists as some form of the creative force from which the human condition arises, we can be assured "he" is not up there behind the Milky Way twiddling his thumbs and wondering what to do with this mess. "He" is here among us, intermingling with every aspect of life, or perhaps even manifesting as all of life. Mystics from the beginning of time, including Jesus Christ himself, have been saying this, and evidence from the contemporary physicists is pouring in.

And the verdict is...

...Not Separate.

We have trained ourselves to believe that the mystery of life is as basic as getting up, making the morning coffee, sending the kids off to school or going to the office, watching TV, having occasional sex with our mate, and going back to sleep. However, at the same time, as science and technology are driving us deeper into the well of consumption and materialism, they are simultaneously revealing discoveries that if considered even in a cursory manner would destroy the very notions of separateness that allow us to indulge in these things. Some scientists say they can prove the existence of God.

Every human being is actually, literally and physically a part of everybody else—inextricably linked with one another not just through a sappy "heart connection," but through our cells, molecules and breath. Energy cannot be created or destroyed, and therefore everything that is here now—both animate and inanimate—has always been here, though in a constant state of flux and change, and will continue to remain long after we are gone.

Human beings are constantly exchanging atoms. You have the highest level of atomic interchange with the people you live with and the people you sleep with (hence married couples begin to look alike, and even think alike), but it is also true that every human being literally contains the atoms of Jesus and the Buddha. Nobody escapes this—even if you live on the top of a mountain, in the purest air, with nobody around for thousands of miles...maybe there is less contamination, but everything is present—matter is matter. Death, therefore, can be understood not as annihilation, but instead as a dissolution and rearranging of form.

How is it that for tens of thousands of years mystics have known and described in detail that which only the most advanced scientific equipment is now able to detect? If we are able to entertain the notion that every molecule within our bodies has existed since the beginning of time and has lived as endless configurations of life forms in the shape of plants, animals, humans and the invisible, it is plausible that knowledge of the whole universe lies within us. A popular saying in Eastern spiritual traditions says, "The microcosm is the macrocosm; the macrocosm is the microcosm." This means that the whole of the universe is contained within the smallest particle. Therefore, although much knowledge is not accessible to ordinary human perception, it is conceivable that a person who dedicates his life to studying the depths of his being can obtain data that only the most highly sophisticated laboratories are able to produce.

Understanding how interconnected we are, we can either embrace the context of touch in order to nourish the greater life we are a part of, or we can insist on our separateness and not take responsibility for the exchanging of life that we are participating in.

The Power of Thought

We touch each other through thought. In everyday terms, interconnectedness means that many people who have the experience of thinking of somebody they haven't seen or spoken to for a long time, will suddenly receive a letter or a phone call from them. The first five or ten times this happens, we call it coincidence; after that, it usually starts to rub us a little more, though we still may attempt to deny it, terrified of our own potential greatness.

Every thought we have (each backed by an emotional charge—be it positive, negative, or neutral) sends out a measurable electromagnetic charge into the atmosphere, and if it is about a specific person or thing, it is projected in that direction. Thoughts are things! They touch other people whether we want them to or not, and have the power to either kill or heal—literally. Thoughts can be "picked up," "read," and "communicated with," depending upon how receptive one is. Even when one is convinced that he or she is dense and insensitive, they are nonetheless impacted by the thoughts of others. When we think our thoughts are a private affair—that just because we are not expressing the hatred or desire we are harboring for a person that he or she does not feel it (whether it is recognized by them consciously or not)—we are fooling ourselves. It is for this reason that if I pass you on the street and you are in a good mood, I feel that from you. Similarly, if you are in a horrible mood and I walk into the room, I also feel that, because you are actually radiating outward what you think is your private mood. This is why we find some people uplifting to be around and others draining.[3]

Educator and author Lalitha Thomas uses the analogy of color to describe the principle of how our thoughts affect one another. She invites us to imagine that each particular thought had a color to match its quality. For example, an angry thought such as, "I'm so pissed off at you for doing that to me!" might be dark red, and an angry thought with malice and revenge such as, "I'm glad you feel rotten, you had it coming to you," may be dark red with black splotches. A lustful, "I want to jump on you right now" thought could be muddy green. "Don't say another word to me," may be gray-brown, and similarly, "I'm really glad to see you," might be lemon yellow.

Thomas further suggests to imagine that every time that we have such thoughts about somebody, that he or she would be splattered in that color paint. Take the example of somebody showing up at your house to do some repair work after you have just finished cleaning the house thoroughly. After giving you a hasty and unfriendly greeting, he proceeds to trample through the freshly vacuumed living room, leaving mud traces across the carpet. Feeling annoyed, thinking to yourself how rude he is, you splatter him with a blob of murky gray paint on his chest and beard. Aware of your annoyance, but unconcerned, as he takes out his tools he lights up a cigarette without asking permission. You think, "How disgusting," and throw a green slimy color on his head that drips down his face. As he starts drilling, he takes out his portable stereo and cranks up his favorite heavy metal music. Your repulsion smears him with dark red. He then hands you a bill, of which fifty percent of the charge is labor. Furious that you have to pay *anything* for having had him in your house, you top off your painting by soiling him thoroughly in rusty brown. By the time he leaves your house, not only does he have his own rudeness, ineptness and insensitivity to live with (which are all manifestations of insecurity, shame and pain), but he is covered with your negativity as well. Furthermore, you are left full of the righteous negativity that has coated your insides with the same colors, and is likely to stay with you throughout the afternoon.

I had been considering this principle one day when my client came in to a session telling me that although it sounded crazy to him, he had been up all night warding off an "attack" from his father, though his father lived across the country. The following week he came in with a hostile and accusatory letter than had been written by his father. The letter was a response to my client's confrontation of his father for having treated him unfairly in a business deal they were in together. The letter was dated early morning of the very night my client had felt attacked (completely synchronous given the several hour time difference from coast to coast). What is commonly referred to as psychic phenomenon is little more than the receptivity to, and exchange of, specific thought-forms in the atmosphere.

Sometimes we don't want to take responsibility for our thoughts and feelings. We often want a person who has hurt us to feel as badly as we do. Sometimes consciously and sometimes unconsciously, we seek revenge on others, telling ourselves that they deserve it. Unfortunately, we suffer this revenge as well. In order for me to throw murky gray, slime green and dark red on somebody else, I must be overflowing with these substances myself. Furthermore, as our thoughts contaminate those around us, those people are going to respond to us from behind these color shells, prompting a ricochet effect that commonly leads to the kind of heated arguments between people that appear to have arisen out of nowhere.

Similarly, our thoughts can heal one another. For example, if your child comes home from school crying, and leaves a trail of mud behind him on the newly vacuumed rug, and instead of criticizing him you open your arms to him thinking, "My sweet child, what could have happened to upset him so?" you cover him with a soft rosy color. As he tells you how bad he feels for getting a poor grade on the spelling test he studied so hard for, you stroke his hair to comfort him and bathe him in light blue. Later you hear him upstairs making a racket while you are quietly reading your book, but instead of yelling at him to stop the noise you think, "He's having a rough night," and you smooth him over with white. By the next morning, he is fine and you have both been nurtured by your warm thoughts.

This is how prayer works. Larry Dossey, medical doctor and internationally renown speaker in the area of spirituality and medicine, has devoted many years to the study of how prayer affects healing. In his book, *Prayer is Good Medicine*, he elucidates that in the more than 130 scientific studies done in the area of healing, over half of these experiments strongly indicate that prayer works.[4] In a study done by cardiologist Randolph Byrd of the University of California General Hospital, groups of very ill people were divided into two groups. Far from the hospital setting, one group was prayed for by an anonymous group of strangers, and the other group was left alone—neither group aware of whom would be prayed for. The effect of prayer on their healing was noteworthy—those prayed for experienced a more rapid and effective recovery.[5]

When somebody lovingly says, "You'll be in my thoughts," if he or she means it, you are in good shape because no matter who they are or where they are, you will be helped. The energy and attention of the other enters into your physical body, exciting the immune system, activating the mental chemistry, and adding the extra elements to it that just may take you over the edge of a healing crisis. There's really no mystery to it.

If individual thoughts holds this much power, it follows that people tend to be joyful and relaxed in cultures where there is an overall feeling of connectedness, a positive attitude, and in which the individual is held in the light of his or her strengths instead of their weaknesses. This is one reason why people tend to be so attracted to tropical paradises such as the Caribbean Islands or Hawaii. Aside from the obvious physical beauty of the places, and although visitors may never interact with the islanders, the environments themselves radiate qualities of relaxation, affection, openness and non-aggression. Whether visitors are aware of it or not, they pick up on this energetic field which is stronger than their own personal field, and find themselves feeling similar qualities of ease and relaxation.

It likewise stands to reason that if a depressed, cynical and addictive person makes a commitment to stay in a nourishing and affectionate environment, he or she will eventually end up, lawfully and naturally, patterning off of the environment in which they have steeped themselves.

Science reports that the body is continually renewing itself. The skin replaces itself entirely every month, the stomach lining every five days, the liver every six weeks and the skeleton every three months. In fact, ninety-eight percent of the atoms in the human body are replaced every year, and the whole human body is renewed every seven years! Still, the body ends up in much the same form. This is in part due to genetics, but also because repetitive and habitual thought-forms keep our physiological systems operating in much the same way, therefore reproducing a similar make up, even if this includes imbalance and disease. In other words, much of what keeps ailments and physical dispositions locked into our bodies is our minds.

The purpose of considering the impact of thought is not to inspire worry, terror or guilt in the individual concerning his or her own thought processes (as it originally did for me when the reality of it first sunk in). Rather, it is meant to help us rid ourselves of the illusion that we are separate, independent and unimportant, and also to bring us face to face with the responsibility we have as a result of our interconnectedness. Mother Teresa once said, "We ourselves feel that what we are doing is just a drop in the ocean. But if that drop was not in the ocean, I think the ocean would be less because of that drop."

When we see the situation of our interconnectedness clearly, there arises within us a responsibility to it. If we don't take responsibility for our vision, we suffer our own conscience. Our interconnectedness reminds us that we are not solid, static, "preservable" beings, and that things are not as they seem to be on the surface. Although this initially may give rise to feelings of insecurity or temporary panic, when we recognize from the bottom of our being that nothing is stable and that nothing can be held onto—when we know through and through that we are a small, but necessary element of a continually changing universe, we cease to need to protect ourselves from annihilation in the same way. We start to show up and take our place in the world, however great or insignificant that place is.

Yet, many people prefer to play ignorant, their claims of "disbelief" serving to cover the reluctance they feel to face this responsibility. However, whether we accept responsibility for our actions or not, their impact is equal. The Eastern notion of *karma* is simply a name for the ever-present law of cause and effect. Whereas many consider karma as something to be feared, equating it with a system of reward or punishment which rests on the premise of a supreme and separate God who sits in judgment of human beings, Tibetan master Jigme Rinpoche suggested that this is a limited and misinterpreted view of karma. He said that when karma is instead considered from the perspective that positive actions and gestures can have helpful and live-giving results, as opposed to viewing karma entirely from the negative, it opens up an extraordinary domain of possibility.[6]

Interconnectedness and the Environment

Nowhere is our responsibility for our interconnectedness more evident than in the relationships that comprise our immediate environment. The urgent sense of responsibility that a mother has for her newborn child dissipates all too quickly, however, as situations arise when it is inconvenient for her to do what is needed; and when self-interest is threatened, many people fail to apply that urgent sense of responsibility to their mates and loved ones. The famous Sufi saying, "Have faith in God, but tie your camels first," could be better paraphrased in the modern day as, "Go to church on Sunday, the PTA meeting on Wednesday, and the movies on Friday night, but first make sure you're hugging your kid and not abusing your husband or wife." Our responsibility is not only to take our loved ones' spoken wishes into consideration, but their unspoken wishes as well.

As with family relationships, our relationship to the earth tends to be one of either separateness and disregard, or one of connectedness and consideration. In present times, everyone is a part of global destruction in a superficial sense, given that anyone who drives a car is participating in an oil industry that not only taxes the earth, but that has violent, competitive and aggressive undertones in every aspect of its functioning, and that the chemically-processed foods that we all eat involve various types of exploitation of the earth in their production. On the other hand, the two hundred sheets of recycled toilet paper we use a week will hardly make a difference in terms of deforestation. However, what does happen when we make the choice to buy the recycled toilet paper, or organic carrots (or whatever other conscious choices we make), is that the positive thoughts that motivate our action go out into the atmosphere in the form of energy, and our intention to regard our environment with consideration is strengthened and purified.

We touch our environment by how we relate to it. Whether it is understood or not, the consequences of our present context in relationship to our environment will have a lasting impact on subtle, often unseen leves. If every aggressive and violent thought carries with it a destructive electromagnetic charge, every abusive act is like an electromagnetic "bullet," and acts of violence against the body of the earth are equivalent to exploding a nuclear bomb on its surface. There is a direct link between our thoughts, our immediate environment, the body of the earth and every action we take. Either we relate to the environment from the context of touch, or we cause indirect harm through our ignorance.

While at the well washing clothes one morning with my indigenous friends, I realized how "out of touch" I was with my environment. Juanita, the woman whose house I lived in, asked me where my family in America got its water. I was stumped, my mind quickly scanned the local reservoirs in my home town, trying to determine from where our water came, but I wound up with little more than a vision of the kitchen faucet. I responded with embarrassment that I didn't know. Surprised by my ignorance, Juanita's mother, Doña Catalliña, who gathered wood every morning for the fire, asked me if I knew where the heat in my house came from. I hadn't a clue. I could not avoid the realization of my complete disconnection with the most basic elements of my immediate environment. I saw that it didn't matter how many letters I had written to Greenpeace in my past. When I grab an apple from the fruit bowl and mindlessly remove the sticker with the apple's "identification number" on it, I am living out my ignorance.

When we stop cooking, we cease to be in touch with our food. When we stop growing it, we are, again, further away. Every step that we distance ourselves from our immediate environment is a step out of relationship and away from integrity. I'm not suggesting that we all go back to farm life, we never will. But there are practical steps we can take. We can go barefoot in our homes or in our yard, feeling the ground beneath us. We can make a small garden and grow a few vegetables, or grow our own herbs in a window box if we live in an apartment. We can learn how to knit a cap, or start a compost pile. To be in touch with our lives, we have to be in touch with all of the elements of it to some degree, however small. A village man once said to me, "At a certain point in everybody's life, they need to build a house of their own, even if they've never picked up a hammer in their life." We may not build a house, but we can build a sandbox or a shelf. All of these things allow us to touch and be touched by our environment. When we have reverence for our environment, we still may eat processed foods, or choose to buy commercial vegetables over organic ones, but we are aware that our actions are costing the environment. Instead of feeling guilty about it, we can simply be honest with ourselves and eat with gratitude—the toxic effect of our actions thereby lessened.

When we know that we are affecting our world, we give up the lie of separation and cynicism. Irina Tweedie says:

> *The Realization that every act, every word, every thought of ours...influences our environment...If we only knew deeply, absolutely, that our smallest act, our smallest thought, has such far-reaching effects...how carefully we would act and speak and think. How precious life would become in its integral oneness. It is wonderful and frightening. The responsibility is terrifying and fascinating in its depth and completeness, containing as it does the perplexing insecurity of being unique and the profound consolation of forming part of the Eternal Undivided Whole.*[7]

IT'S JUST LIKE THE BUDDHA SAID

The Buddha, who was well aware of the principle of interconnectedness, could be considered to be one of the greatest psychologists of all time. Two thousand five hundred years ago, after having sat under the famous Bodhi tree for seven years, he proclaimed to his disciples that, "All life is suffering." The Buddha, who was the most compassionate of humans, and who also represents that place of perfect clarity and acceptance that exists within each of us, was not being cynical or pessimistic as is so often misinterpreted. Instead, he was speaking to the reality of our condition. He wanted to teach people what life is, so that they would have the opportunity to come to terms with it. He was letting people know that the fact that their minds were tormented, that they had such difficulty with their relationships, with food, with how to be a good parent, with feelings of rage and depression—did not mean that they were in any way "bad" or doing anything "wrong." He was asking people to look at the human condition, to consider the ways that we both hurt others and are hurt by them, to acknowledge our fear of sickness and death, and to realize that all of this is included in the package of being alive. This was the first of what are referred to as *The Four Noble Truths.*

Next, the Buddha explained to those around him that there is a cause for this suffering. He called it thirst, or craving. He was not only referring to the desire for, and attachment to, sense pleasures, wealth and power, but also the attachment to the masses of ideals and ideas, points of view, concepts, theories and beliefs that constitute the relentless chatter of the mind—the chatter that is constantly telling us what is O.K., what is not, what we want to eat, what we do not, whom we like, who will hurt us, why we're no good, why we're so great, etcetera! This chattering mind is the part of us that did not bond (discussed previously in the book); the part that labors under the illusion of separation, that is always looking for something to fill an obscure and unnamed void within. It is the part of us that is terrified of the insecurity that is a fact of our lives, and instead has created fantasies about reality that, even if they are inaccurate or bring further suffering upon us, we would rather hang onto than face the vulnerability that is at the base of our humanness.

Much to the relief of those who sat by him, who were rapidly becoming depressed as he spoke, the Buddha then set forth *The Third Noble Truth*—that there is emancipation, liberation, freedom from this suffering, and that it is not found in a pint of Haagën Däz™ or in front of the television set. He explained that whereas it is impossible to escape the realities inherent in a human existence (such as illness, aging, death, and sorrow), it is possible to cease to allow these forces to run our life. In choosing to see clearly and accept the reality of our lives (realizing that whatever happened in our childhoods is no longer happening yet cannot be undone), accepting that what is *is*—we stop wanting things to be other than they are, and in doing so cease to live in fear and reactivity to the things that we cannot change, accepting that we cannot control our lives. He was telling the group that we are not victims of our lives—that there is a way we can learn to move in and out of our days with kindness and clarity.

Lastly, Buddha explained a systematic method that would enable people to find their way out of suffering, which he called *The Eightfold Path*. This consisted of: Right Understanding, Right Thought, Right Speech, Right Action, Right Livelihood, Right Effort, Right Mindfulness, and Right Concentration. Although there are fine points to each of these eight components, he was basically describing how human beings would naturally come to live if they were in touch with themselves. His way was called The Middle Path because it did not advocate seeking fulfillment through an indulgence of the senses, as is so common in the West, nor did it suggest a denial of the body or the senses. Instead, we passionately seek fulfillment in the ordinary world—the wonderful, the terrible and the boring.

I am not Buddhist, and am not suggesting that the only way to get in touch with ourselves is to sit under the magnolia tree on our front lawn for seven years. For parents with young children, or individuals who work forty hours a week, it is simply not plausible to make such extreme gestures. On the other hand, emancipation from our suffering, and the cultivation of an inner freedom that would allow our lives to be full of touch and intimacy, does not come easy. It requires a ruthless and honest examination of our present state of touch-starvation and touch-phobia, and the willingness to *act* rightly, again and again, until our lives evidence greater touch and intimacy.

Buddhist cosmology suggests that of the many realms of life that a human being can be born into—including hell realms, heaven realms, and realms of the gods—that the greatest opportunity exists in being born into the human realm, because here there is always a possibility to transform, to change course, to be different, to get in touch with who we really are.

The way out of this suffering is by beginning to get in touch with ourselves and with others. Getting in touch needn't be dramatic, but can begin in simple ways. You start by taking ten full seconds to give your wife your undivided attention while hugging her when you (or she) come home from work at the end of the day. You begin when, instead of saying "fuck you" (either silently or to her face) when your close friend snaps at you, you turn to her and touching her arm ask, "Are you having a hard day?" You catch yourself entertaining negative thoughts about somebody and instead turn your attention to what you are doing. The Buddha didn't have it easy, and neither will you, but every step you take is one step closer.

TOUCH IN DAILY LIFE

> ***It is no use walking anywhere to preach unless our walking is our preaching.***
>
> –St. Francis of Assisi

It is possible to become a student of touch, to aspire to live one's life in the context of touch, and to take practical steps in this direction. The earnest student gets glimpses of this context, and is increasingly able to recognize and observe it in others when it is present, thereby developing a refined sense about how this can show up in his day-to-day life. When he is inspired by the context of touch he makes choices about how to lead his life, and uses discipline to follow through with this inspiration when he is not feeling it. While one cannot "make" true touch happen, he or she can act according to its principles. In other words, until the mood of touch comes on its own, one can keep touching regardless, ceasing to feel disempowered by the absence of it when it is not present, and living on the premise of the truth of it while waiting for it to surface within more fully. This next section considers pragmatic ways that touch can be practiced in the context of one's daily life, beginning with an overview of some specific ways the individual can get in touch with himself internally, then moving into a discussion about how one can touch others by extending his or her hospitality to them, and finally by a consideration of service, which is perhaps the highest form of touch.

Touch on the Inside—Breath, Meditation and Martial Arts

We can have no context for touch in our lives if we are not in touch with ourselves. I recently spoke with a young man named Ray who had psoriasis—a condition of the skin that results in times of extreme irritation and long periods in which the individual cannot stand to be touched in any way. Although Ray was aware that psoriasis often has a psychosomatic component (meaning that it originates in the mind and manifests in the body), this did nothing to ease the feelings of intense isolation and loneliness that ensued from the lack of physical contact in his life.

Ray recognized that the traditional psychotherapy he had been participating in for the past two years was unable to reach the places where he was most hungry for contact, so he turned to meditation. As with most forms of meditation, his practice required a great deal of patience. However, through this process, Ray found that he was literally able to *touch* places within himself that he had been previously unable to access. Through breath and awareness, he could move into both wounds and pleasures that were hidden deep within his body and mind, and in doing so he discovered a strong sense of intimacy and relationship within himself. This was simultaneously accompanied by a remission of his rashes and irritation.

Unless we are able to feel inside of our own bodies, we cannot be in relationship with our surroundings. If we are not in relationship with ourselves, we do not feel at home in the world, as if we had somehow been abandoned and are now in a game of tug-of-war with the universe. Yet the body knows all, and we can use meditation to optimize our receptivity to the constant feedback and messages that the body provides. Meditation and other forms of internal practice help us polish and refine the receptors so that the messages we get are clear and accurate.

Furthermore, when we are sensitive to our inner life, we know when we are "off" or have gone astray. When people say, "I'm in such a fog," or, "My body feels like lead," they are expressing a blockage in the mind-body connection that is inhibiting their receptive capacities. Internal practice is yet another avenue to increased consciousness and awareness.

Years ago when I was leaving for a ten-day meditation retreat in late December, my mother called and asked how I planned to spend my holidays that year. When I told her what I was doing, her response was, "Why don't you just stare out the window for ten days?!" I laughed, and explained to her that although meditation practices are often very simple, they involve a subtle process of working with the mind and body such that we cease to be slaves to our minds, thus freeing up the natural energy flow of the body. My point was that although these practices may look like just sitting on a cushion (as meditation does), or doing very slow and intentional movements (as is often the case in the martial arts), when the wisdom of the body is invoked by practice, it catalyzes certain processes that cannot be compared to gazing out the window!

Bear in mind that none of these practices are meant to replace the need for basic touch and affection in relationship to those in our environment. It is worth noting, by way of precaution, that countless individuals in all religions have, in their terror of intimacy, turned to lives of celibacy and meditation in order to disguise, even to themselves, their fear of touch and their fear of others. In an ideal situation, touch of self and touch with others would optimally build upon and enhance one another, creating an overall deepening of intimacy based on the individual's increased sensitivity.

Breath

Of the breath, Frederick Leboyer says, "To live freely is to breath freely." Called *prana* in Sanskrit, *ruach* in Hebrew, and *pneuma* in Greek, it is the same breath, the same life force, that is constantly exchanging life with the environment around us—breathing out carbon dioxide to feed all plant life, and taking in the oxygen that it provides. (To think we already knew this in third grade!) The breath has tremendous healing properties which have been known to all spiritual traditions, and which are now being discussed routinely by medical doctors. For not only is the breath directly connected with states of anxiety and/or relaxation, but through the breath we can directly control our blood pressure, body temperature, heart rate and so forth. I guarantee that if you stop right now and take a few deep breaths that you will instantly feel differently—probably more relaxed, centered and present.

Breath also connects us to our feelings. You will notice in a person who appears to be "in his head," or who "thinks too much," that his breath is very shallow. Children who have been traumatized in any way will often automatically develop unnatural and constricted patterns of breathing in order to ward off the intensity of their underlying emotional pain. These children grow up to become anxious adults who are probably entirely unaware that their natural flow of breath is obstructed, much less that their breathing patterns are making their physical and emotional lives extremely difficult. When we breathe we enter into relationship with the body, the center of all feeling—both pleasurable and painful. Through breath alone we can access states of deep calmness, clarity and equanimity, as well as move through painful emotional blockages.

If you want to learn more about your own breathing cycles, you can try the following:

An Exercise in Breath Awareness:

- Lie down on your back and try to draw your next breath all the way down to your abdomen. Notice how far the breath goes in. For many people, it does not go past the lower throat or upper lungs. It is important to refrain from judging yourself in any way, as to do so will prompt you to try to control the breath, making it how you think it should be instead of using the opportunity to ascertain information about your breathing cycles.
- Next, take a moment to place relaxed attention on your abdomen, just watching it rise and fall. Do nothing but observe.
- Now, try to breathe down into the lungs, the diaphragm, the abdomen and belly—naturally, simply, easily. Pay attention to the rhythmic pattern of breathing—again, not judging it, just observing. Notice whether it is easier to breath in, or to breathe out? Is the in-breath longer, shorter, or equal in length to the outbreath?
- As you become more relaxed, notice if there are particular places of constriction or blockage, or if the breath flows smooth and freely. You can also experiment with what it feels like to inhale into various parts of the body—the legs, the head, the hands and so forth, and what it feels like to intentionally block

the throat, or to not breathe fully. You can experiment with holding your breath or trying to relax it. The point of the exercise is simply to bring your awareness to how breath moves in your body.

Practicing with the breath needn't be a big production. Many of the fancy postures and closed-eye, dramatic breathing exercises that have become so popular are little more than cosmic fluff. You can renew yourself through breath in the middle of a hectic day at the office and nobody will ever notice; or you can do it while driving in a car—taking a moment to experience one full inhalation, and to let out one full exhalation.

It is helpful to be aware of your breath throughout the day—when you are eating, when you are hurrying, when you are relaxed. It is particularly useful to pay attention to the breath as you pass through the various stages of falling asleep and waking up (when not jolted by your alarm clock), for at these times the body is usually more relaxed, breathing according to its own organic rhythms. From that experience you can gain a reference point in yourself as to what it would mean to breathe freely. Eventually you will develop a more conscious relationship with your breath.

Meditation

There are many forms of meditation—some aim to focus the mind and sharpen attention, some are a type of prayer, others are solely intended to bring about peace and relaxation. Regardless, they all share in common the aim of helping the individual to get more in touch with his or her inner self, to see behind the screens of the superficial.

I will not give meditation instructions here. The subject of meditation is vast and complex, and would require a book of its own. Depending upon one's make-up, various practices are more or less suitable to the individual's needs. However, if meditation is a subject that interests you, one that you wish to pursue further, and you don't already have your own practice, there are some useful guidelines to keep in mind.

The New Age scene advertises an endless list of possibilities for "Instant Enlightenment," "Spiritual Awakening in Three Days at a Fifty Percent Discount," "Everlasting Peace of Mind for only $175," "Soul Retrieval," and on and on. The propaganda can be enticing and convincing, and for this reason it is helpful to be wary and to use discernment when choosing a practice. First of all, it is important to understand that meditation is not equivalent to psychic healing, psychotherapy, hatha yoga, body work or deep relaxation. Secondly, it is pointless, and even dangerous, for beginning meditators to engage in advanced practices. Any promises of instant anything in terms of meditation should be considered as a warning flag. Thirdly, learning meditation should not be a financially costly process. Many times it will be offered free of charge or by donation, or if you are learning in a retreat setting, the costs will generally not be excessive.

Many basic types of meditation are grounded, reliable and effective. Among these are Vipassana (also called Insight), Zen and Tibetan Buddhist meditation practices and Centering Prayer, which are taught in many larger and smaller cities and towns. These forms of meditation generally consist of sitting on a cushion for thirty minutes to an hour, and using the breath or another point of focus in order to undertake a process of self-examination. The "success" of meditation practice depends primarily upon patience, persistence, dedication and consistency. When we recognize the degree to which we are out of touch with ourselves, we will come to appreciate that getting in touch with ourselves is a slow process that will occur gradually. Other types of spiritual practice can bring us into specifically defined aspects of development and growth, but grounding ourselves in a simple and safe meditation practice will give us discernment and a more clear picture of what lies beyond.

Martial Arts and Yoga

There are numerous types of martial arts and yogas. Some, like *aikido* and *judo,* involve direct relationship with others, and others, like *hatha yoga* and *chi kung,* are primarily practiced alone. Most forms of martial arts and yoga involve a relationship with a teacher, who can be a source of valuable feedback, as the subtleties and distinctions in body posture, breath, attention and attitude are important. The intention of martial arts and yoga is to activate the life force, the *chi*, the natural energy of the body, and like the other practices listed here, to aid the individual in getting in touch with the inner workings of his or her body, and to bring him or her into a more direct relationship with his or her environment.

Though commonly thought of as a means of self-defense, the underlying goal of martial arts training is the achievement of mental, physical and spiritual strength. Aikido master Morihei Ueshiba asserts that, "Aikido is the study of the spirit."[8] Although what is visible to the human eye is a series of physical techniques, such techniques are merely vehicles used to express spiritual principles that have deep roots in Eastern traditions.

Martial artists who study not only technique but who apprentice to the context of the form they study, become deeply intimate with the subtle energies of the body. When an individual is in touch with himself in this way, he becomes a wellspring of knowledge as to the necessity and practical applications of this touch in his own life.

The spirit of yoga is similar to that of the martial arts. The numerous somatic exercises consisting of postures and stretches (known as *asanas*) that are usually thought of as yoga itself, are actually a form of *hatha yoga,* and are one aspect of a greater spiritual discipline that is designed to bring an individual into a harmonious relationship with life itself. Yogic exercises not only bring awareness and strength to the body, but are a form of self-touch. The practitioner of yoga knows her own body, and because of this sensitivity is more "in touch" with her surroundings.

The other principle forms of yoga (which literally translates as "union") are *karma yoga,* the path to union through selfless service, and *bhakti yoga,* the path of union by way of devotion (to God, Jesus, Allah, etc., or to an awakened spiritual teacher). The following sections on hospitality and service could be considered as forms of karma yoga. Hospitality and service could also be considered expressions of basic goodness, integrity and the only sensible outcome of a life lived in touch! Essentially, all touch is about devotion to The One—the one Love, the one Truth,—the One Whatever! We touch because we love. We love because we have been touched.

Hospitality as Touch

> *Hospitality is the greatest law given to man. If he knew how to obey this one law he could overcome his imperfections.*[9]
>
> - E.J. Gold

The Hindu tradition in India proclaims that God lives within every being. When a guest comes to one's home, he or she is to be treated as though God has entered the house. By hosting their guest in this manner, the Hindu people practice reverence and consideration of all beings, irrespective of their personal biases and preferences. Like the mother who—whether she feels angry, annoyed, or adoring in relationship to her children—continually provides for them based upon her underlying love and commitment, so the host places the guest before himself, and in doing so offers himself to service of the other. Hospitality is clearly a subset of service, but is considered separately here to give it a special seat of honor among the many forms of service.

In Japan, the Tea Ceremony is a highly respected and refined practice of hospitality. In mastering the simple act of serving tea and receiving it with gratitude, a synthesis of the highest cultural ideals, including religion, morality, aesthetics, philosophy, discipline and social relationship, is expressed. In the offering of tea, the host offers the guest attention, presence and humility—exemplifying an optimum moment of contact between two people.

All traditions at one time were rooted in the recognition of hospitality, and many individuals and traditions still practice this hospitality very intentionally. But the faster we continue to move as a culture, the more fast food pick-up replaces dinner-table gatherings, and evenings on the Internet or in front of the television are substituted for quiet family time, hospitality and gestures of personal consideration fall rapidly to the wayside.

Whereas the subtleties that give hospitality its particular elegance show up in the details, the details are not the point. Hospitality may include offering food to a guest, preparing his or her room, initiating conversation at times and allowing for quiet at other times. It may be the extra care that was taken to dust thoroughly, to put a vase of flowers on the table, to make the bed in a flawless manner, but all of these "details" are actually ways of honoring the person before you. By placing positive attention on your guest, you allow him or her to feel cared for and nourished, making yourself available to him or her, and communicating a sense of welcome. When your guest feels touched, it is not *because* of the fresh sheets or the flowers, but because of the intention and care that lies behind these gestures. He is responding to the mood created by his host's attention.

Hospitality is touch. It is an external gesture of respect and regard that does not demand anything of the other, thereby allowing him or her the full space to receive the gesture of touch that is being offered. When you allow your guest space, while still being attentive to her, you are giving her the space to bring herself forward, and as a result will often receive the best of her.

Many people are extremely shut down, and physical touch is just too threatening to them. Another unique aspect of hospitality as touch is that it gives a person who is fearful of touch an opportunity to feel nourished in ways that they may not ordinarily allow themselves to receive. To these people we offer ourselves more subtly and indirectly. Sometimes people send greeting cards, love letters or gifts to this end. The warmth extended by such hospitality may provide just enough safety for the other to let the touch in. In a touch-starved world, small gestures mean a lot.

When we intentionally extend our hospitality to somebody else, we are able to experience our own respect and regard for our guest. When we are carefully setting up a room for a guest, or preparing a special meal for him or her, or even making a cup of tea or offering them a comfortable chair, we often experience within ourselves qualities of nurturance and a genuine desire for the other's well being, and in doing so are touched by our own goodness. Hospitality is a practice of service to all beings, in spite of our biases and judgments. Remember, the paradox of touch is that the one who serves, receives the most. Even if the host ends up sleeping on the less comfortable bed, or eating the leftovers, if they have served wholeheartedly, they will reap the fruits of their own giving.

Guests need to practice hospitality too. It is as important to be a good guest as it is to be a good host. On the surface, this may show up as cleanliness, appreciation and a gracious acceptance of whatever is given. But more importantly, the guest's receptivity allows the giver to give—to pour forth themselves into an open receptacle. When both host and guest are participating in their respective roles, hospitality becomes a form of reciprocal relationship. Gratitude for giving and gratitude for receiving come together to produce a mood of harmony and communion between the two.

I learned the hard way. Though the principle of hospitality is an extension of common sense and basic integrity, I had not yet understood it clearly at eighteen when I went to work in Dzidzilché, Mexico. To express the appreciation for the work my colleagues and I were doing in the village, a very poor family invited us over for a special dinner. When they proudly presented the chicken they had slaughtered freshly that day, my stomach turned over. Actually, it was my mind that probably turned first. I had been a strict vegetarian for some years at that point, and the idea of breaking my vow was devastating. I politely refused the chicken, and responded the best I could to their baffled questions about why I would not eat meat. No matter how clear my responses, I could not avoid the look of disappointment on their faces. Several months afterward, when I was able to gain perspective on the situation, I saw how inappropriately I had responded to the situation, and felt remorse because of it. The principles of gratitude and receptivity that underlie hospitality are far more important than any rigid morality concerning personal preferences.

We can also offer hospitality to ourselves. We do this by taking an attitude of nurturance and consideration toward ourselves (which is distinct from selfishness). We keep our personal spaces neat, our clothes clean, our bodies fed with nourishing food. When we communicate an attitude of self-care inwardly, we feel respected and cared for in the same way we wish our guests to feel. We locate a core of self-respect within, and are more free to allow intimacy to flow in our lives. Of this, Father Henri Nouwen, author of over thirty books and mentor to thousands of seminary students says, "When we have found the anchor places for our lives in our own center, we can be free to let others enter into the space created for them."[10]

Service: The Touch of God in The Flesh

> *I don't know what your destiny will be, but one thing I do know: the only ones among you who will be really happy are those who have sought and found how to serve.*
>
> –Albert Schweitzer

My grandfather used to tell me the famous story of a man who was drowning in the ocean and who turned to God for help. He called to God, "I am drowning and I need your help! I am a man of great faith and good deeds so you must save me!" The man waited, growing tired from treading water, but confident God would come at any moment because he was pious. Meanwhile, a fisherman came by and saw that the man was drowning, "Here! Let me help you," said the fisherman throwing him a life preserver from the boat. Sputtering for breath, the man answered, "No. I am waiting for God to save me, for I am a man of great faith." The fisherman tried to insist, but the drowning man would not hear of it. Shortly after, a water-skier happened to come by, and seeing the condition of the man, offered him his rope so that he could be towed back to safety. Again, the man stubbornly refused. The fisherman must have gone to shore and called the coastal police, because soon there was a helicopter above that dropped down a ladder. A voice called through a megaphone for the man to grab hold of the ladder. Though bloated and nearly dead by now, the man refused, calling up to the pilot through the roaring ocean, "I have faith. God will save me."

The man drowned, but apparently because of his faith was taken to the gates of heaven where God was waiting. The man shook his fist at God, "Why didn't you save me?! I am a man of great faith and merit." Half-laughing, God responded, "What are you talking about? First I sent you a fisherman. When you stubbornly refused his help I sent you a water skier, and I even went to the trouble of sending a helicopter!"

If God or Spirit—in whatever form we imagine it to be—is going to touch us with help and grace, it is going to happen here and now, through the people in our immediate vicinity. As was discussed earlier in the chapter, God does not live in the sky, and if we don't find Him (Her, It) in one another, nobody is going to be waiting for us at the pearly gates of some fairytale heaven. A few individuals may be able to experience the touch of a formless God, but for the rest of us, it's either each other or nobody.

Jesus and Buddha are Joe and Gertrude who live on the corner. The only thing that makes a saint a saint is that they have given their life over wholly to the service of all humankind. Saints don't just tithe ten percent of their income to the church, or work in a residential treatment center, or serve as the PTA president, although they may do any of those things. The saint *lives* for others. Any powers they have acquired are insignificant manifestations of their service, and are only considered to be of value if they can be used to benefit others. Whereas it is unreasonable to assume that the average person can find within themselves the willingness to serve others to this same degree, the depth of collective wounding that humanity as a whole is experiencing is also unreasonable, and what it will take to remedy the present situation is far beyond what our rational minds care to consider. In other words, even if we are not Mother Mary or Mahatma Gandhi, it is not beyond our capacity in the least to touch others in large and small ways. Mother Teresa has said, "Holiness is not a luxury but a simple duty."[11]

Still, the lives of the saints serve to inspire us, particularly when we recognize their humanness—their deep concern, the pain they feel for others' suffering, their own weaknesses that they must continually confront in order to express a deeper wisdom, and the way they persist against all odds because of the love in their hearts. Appreciating their extraordinary ordinariness, we can then model our lives after those individuals who, in Emily Dickinson's words, "dwell in possibility," instead of looking to the models provided to us by a dying culture.

A striking illustration of this comes from the life of the present day saint Indian saint Mata Amritanandamayi (known by her disciples as "Ammachi"). A man named Dattan showed up at her temple one day. He had been inflicted by a severe case of leprosy, which often eats away at the limbs and skin of the infected individual, resulting in deformity. Upon the man's arrival, Ammachi's close disciples tried to get him to leave the building, and to keep him away from their teacher, fearful that he would infect her. But when Ammachi saw him, she immediately called him over, took him into her arms and hugged and kissed him. Much to the disapproval of those around her, she then proceeded to suck the pus out of his open wounds with her own mouth. In response to the shock of those around her, she said, "Mother doesn't see his external body. She sees only his heart. I cannot discard him. He is my son and I am his mother. Can a mother abandon her son?"[12]

Another extraordinary story about how we touch one another through service is of Russian Orthodox Saint Mother Maria Skobtsova who, after years of protesting the Nazi movement and aiding its victims, was then sent to the Ravensbruck concentration camp. For the two years of her confinement she gave large portions of her food to others, and spent her time boosting the morale and faith of other inmates. Finally, she voluntarily joined a group of anguished older French women who were condemned to die, hoping to inspire them to meet their fate with fearlessness and faith in God, as she did. Ironically, the day after the murder, all the French women prisoners of Ravensbruck were liberated through the auspices of the Red Cross. Mother Maria is quoted to have said, "The way to God lies through love of people; there is no other way."[13]

Sister Chan Khong, a Vietnamese nun who was exiled from her home country because of her political involvement in peace activities, describes her way of touching people:

> *Being a nun in the West, I do not carry undernourished babies in my arms, but teenagers and adults do cry silently as they share the stories of their childhood of sadness and abuse. By listening attentively to their pain and helping them renew themselves, I am able to help heal many of these wounded "children," and this is very close to my ideal of holding the village children in my arms. I am grateful to be able to help in this way...together on the path of love, we can try to make a small difference in someone's life. What else is there to do?*[14]

When Mahatma Gandhi was struck in the chest with a bullet, the last words he uttered as he fell to the ground were, "Ram, Ram"—a Hindu name for God. Saints known and unknown of every tradition are alive and performing their miracles of service daily.

Then there are the ordinary people, who it is unnecessary to call saints, but who may be secretly revered as such by their grandchildren; people who touch others deeply by rendering their services anonymously, and those who, after a life of crime and selfishness, perform one saintly act. Of these individuals Victor Frankl says, "We who lived in concentration camps can remember the men who walked through the huts comforting others, giving away their last piece of bread...they offer sufficient proof that everything can be taken away from a man but one thing: the last of the human freedoms—to choose one's attitude in any given set of circumstances. To choose one's own way."

The woman in the following story is such a person:

> *We had just searched a small village that had been suspected of harboring Viet Cong,* began one Vietnam vet. *We really tore the place up...just up the trail*

> *from the village we were ambushed. I got hit and I don't remember anything more until I woke up with a very old Vietnamese woman leaning over me. Before I passed out again I remembered seeing her in the village we had just destroyed, and I knew I was going to die. When I woke again, the hole in my left side had been cleaned and bandaged, and the woman was leaning over me again offering me a cup of warm tea. As I was drinking the tea and wondering why I was still alive, a helicopter landed nearby to take me back. The woman quietly got up and disappeared down the trail.*[15]

The young can touch through service as well, and they do. Erin, an eighth grader, shared the following:

> *I was going down the stairs after first period. Everyone was walking all over a fifth grader on the ground. I decided to help him pick up his books and when I was finished I told him I'm sorry everyone was stepping on him. I walked away feeling good about myself but I knew he was crying. I know how he felt. I was also a fifth grader once. I'm not telling you this so you think I'm a nice person. I'm telling you this because what I did made me feel good about myself. I hope you take that into consideration. I want other people to know that by helping one person it makes you feel like you helped the whole world.*[16]

A friend of mine corresponded for years with a man on death row who had been convicted of multiple murders. In the last years of his life, the prison inmate was introduced to Mahatma Gandhi's theories of *ahimsa,* or non-violence, and was deeply moved by them. Though there was nothing he could do to save his own life, he began to teach these notions to the younger prisoners, inspiring them to change their attitudes. Before his execution, he had created a non-violent sector of the prison as well as many study groups.

Again, when we embrace the context of touch in our lives, our actions are felt by others *as* touch, whether or not there is any physical contact. In his book *Grace and Grit*, renowned author Ken Wilber beautifully describes a relationship based on the context of touch. His wife, Treya, was diagnosed with cancer ten days after their marriage and died five years later. Recounting their relationship in the final days of Treya's life, Ken said, "In the last six months of her life, it was as if Treya and I went into spiritual overdrive for each other, serving each other in every way that we could. I finally quit the bitching and moaning... just dropped all that. I had absolutely no regrets; I had only gratitude for her presence, and for the extraordinary grace of serving her...it was a profound choice...we simply and directly served each other, exchanging self for other, and *therefore* glimpsing that eternal Spirit which transcends both self and other, both 'me' and 'mine.'"[17]

Sometimes it takes confrontation with our own death or the death of a loved one to bring us to the touch of selfless service, but this "exchanging self for other," can happen at any time of any day, and is our only hope for true happiness.

We all know examples of how we are touched through service in our own lives—they happen all the time, and this is because service is constantly taking place in the realm of the ordinary. One of my fondest memories of my time in San Francisco occurred one day when I was walking to work. I passed my neighbor, whom I had never spoken to though we had lived alongside one another for many months. He was washing his car and in a good mood, and looking at my filthy white car I laughingly joked, "Hey, when you finish yours maybe you can do mine." When I came home that evening and saw my car brilliantly clean, my heart literally fell open. And it wasn't the car—it was my heart that my neighbor had cleaned!

There is a story of a simple village woman who went to a famous sage lamenting that she could not keep her attention on God, and that she feared she would not progress spiritually. The sage then asked her what it was that she loved the most, and the old woman told him that it was her young granddaughter whom she adored more then anything else. The sage then told her to forget about God altogether, and to serve her granddaughter wholly. The woman did so and rumor has it that she became enlightened sometime thereafter.

It is hard to want to live in a context of touch in a world that needs more than one could ever offer, a world that is ever anxious to take *everything* from whomever is willing to give. We fear being swallowed up in another person's needs; we fear the pain we will feel if we open up to their pain; we are concerned that everything we have will be drained from us, leaving us desolate and impoverished.

All of this being true, we are called to touch and to give anyway. In India I would often give money to a group of beggars, and instead of feeling good I would be left distraught by the anger and hostility hurled at me by those to whom I did not give. Still, I gave when I could. Taking just doesn't work. An attitude of scarcity doesn't work. Selfishness doesn't work.

If we have given a cursory glance about us, looking even a fraction of an inch beneath the superficial level of things, we have seen that the only choice is to touch and to give of ourselves. Still, we cannot, nor should we, force ourselves to give. As will be discussed in the section entitled *Heartbreak,* when we acknowledge the pain in our own hearts, this recognition invites compassion into our lives and becomes the starting point of our service.

When we touch others through service, we tend to think of all the benefits *they* will receive from our help, failing to realize that we ourselves are likely to the primary beneficiaries of our service. The saint Swami Vivekananda suggests that we owe a debt of gratitude to those who allow us to serve them:

> *We must remember that it is a privilege to help others. Do not stand on a high pedestal and take five cents in your hand and say, "Here, my poor man!" But be grateful that the poor man is there, so that by making a gift to him you are able to help yourself. It is not the receiver who is blessed, but it is the giver. Be thankful that you are allowed to exercise your power of benevolence and mercy in the world.*[18]

INTENTION

Intention is a remaining piece in this great puzzle of touch that we have been carefully assembling, and is perhaps the most valuable tool that we as human beings have at our disposal. Intention is what turns a touch-starved existence into a life full of touch and intimacy. Intention is not just saying, "Oh, I'd *like* to be one of those people who is really able to touch others." While that wish is a fine place to begin, if we really want to use intention to our advantage, much more will be involved. Intention means defining a highly specified goal and then cultivating within ourselves a complete and thorough desire to achieve this—a full-bodied willingness to do whatever it takes to manifest our particular aim. Finally, we follow through with our intention by way of disciplined and consistent action.

An intention of the degree of clarity and strength that I am suggesting here doesn't usually arise overnight, though it can. Oftentimes, an individual consciously wants to do something—to start being more affectionate with his children, for example—but the unconscious forces within him are not aligned with this wish, since he is still receiving some kind of payoff from his habit (in this case, protecting himself from the pain that much intimacy might reveal), even if it is a destructive one.

Sometimes a clear intention arises from crisis or trauma. For example, an alcoholic man who frequently drives his car while drunk may cultivate a powerful and clear intention to stop drinking when he nearly kills his wife in an accident on the way home from a party. Or an ordinarily cold and distant woman may decide to create more touch and intimacy in her life when, during an extended stay in the hospital, nobody comes to visit her because she has not acted loving toward others in her life. Sometimes intention comes from seeing the price that living in a touch-starved nation continually extracts from our wholeness, such as the individual who is diagnosed with a terminal illness and realizes that he has never really *lived* his life.

Whereas we cannot force intention to work for us, it does become stronger each time we reassert it and refine it. Restating one's aim doesn't have to be done on fancy paper in the presence of several witnesses—intention is grounded in one's relationship with his own true self (or God if you prefer). Intention is a force of unfathomable power, with a life of its own. Once it has been directed, it is as if the goal has already been achieved, just as when a coin is dropped from a tall building, it is only a matter of time before it hits the ground!

The power of intention is illustrated vividly in an account I once heard told by Jack Kornfield, a Jewish American Buddhist teacher, about one of the monks whom he studied with in the forest monasteries of Thailand.

When one enters a monastery—be it Christian, Buddhist, or otherwise—there is a firm decree that one must obey the teacher at all costs in order to optimally progress spiritually. Therefore, when the teacher of a young English monk told him to return to England to start his own monastery, to continue to wear his monk's robe, keep his head shaved and to practice the form of beggary they had done for so many years in Thailand, the monk was determined to obey, irrespective of inevitable feelings of insecurity and uneasiness. As the story goes, this man went walking through one of the many elegant parks in downtown London in his orange robes and with his begging bowl in hand. A curious and unsuspecting jogger stopped when he saw this odd sight, and asked the man what on earth he was doing. The young monk then explained that he was practicing the art of begging he had learned while living in the forest monasteries of Thailand—that he had been sent to England to set up his own monastery under the direction of his teacher. The jogger responded, "Well it so happens that I have a forest which I have been wanting to donate to a good cause. Would you like to have this land to build your monastery on?" Just as the power of intention, against all odds, won this man a beautiful piece of land upon which now stands a thriving Buddhist center, so will it bring genuine affection into our own lives and the lives of those whom we love.

As human beings, we need to become aware of our own power, so that we can consciously harness it as a resource and utilize it to benefit others, instead of remaining helpless at the mercy of pervasive, but relatively unimportant psychological issues. When we turn our attention and power toward getting in touch with our lives, or toward making sure that our children are given ample affection and attention in their early years, or toward acting out of compassion instead of defensiveness, then we have taken hold of the reigns of this force. Until we have admitted our own individual woundedness, as well as humanity's shared ailment, and allowed this to break us open to the compassion that lies beneath, we will be uncontrollably motivated to continue to exploit our own power for selfish gain. It cannot be otherwise. Yet, when we finally allow our heart to be shared with those around us, we will be inwardly moved to sacrifice our own desires for the benefit of the greater good.

The importance of harnessing the resource of our own power as human beings cannot be stated more eloquently than Nelson Mandela did in his inaugural speech:

> *Our deepest fear is not that we are inadequate. Our deepest fear is that we are powerful beyond measure. It is our light, not our darkness that most frightens us. We ask ourselves, "Who am I to be brilliant, gorgeous, talented, fabulous?" Actually, who are you not to be? You are a child of God. Your playing small doesn't serve the world. There's nothing enlightened about shrinking so that other people won't feel insecure around you. We were born to make manifest the glory of God that is within us. It's not just in some of us. It's in everyone, and, as we let our light shine, we unconsciously give other people permission to do the same. As we are liberated from our own fear, our presence automatically liberates others.*

SACRIFICE

Through personal sacrifice we connect ourselves to others and to the greater whole. The individual who leads a life of sacrifice is constantly touching others and the surrounding environment because he is attentive to, and resonant with, needs greater than his own. Contemporary spiritual teacher E.J. Gold says, "Nothing can be accomplished without sacrifice."[18] But, sacrifice is hard—it means that we willfully give up something we personally hold dear (like time, energy, privacy, money) in order to serve the greater good. While our conscience may speak to us, our mind and the voices of the world at large may say the opposite, making it difficult to do what we know we must.

Practically speaking, if we really want to be part of the solution, we have to be willing to make constant, daily sacrifices—taking the opportunities for sacrifice that are presented to us all day long. We would give up having two and three cars per family, and driving to work alone. We would give up excessive consumerism and buying things we don't need and will never use. We would give up investing in companies that we are aware exploit human and natural resources. Certainly we wouldn't need to give up all our pleasures, but we would give up the continuous indulgence in those things that are taxing to the environment.

Yet sacrifice runs still deeper. If we are serious about touching those around us, we would give up our alcoholism, our over-indulgence in television, our gambling, or whatever addictions keep us so obsessed that we become blind to the needs of others. If we are truly committed to touching ourselves in a way that promotes life, rather than death, we would give up eating microwave-radiated food, and taking in high amounts of sugar and caffeine that cloud our minds. We would give up our numbing agents one by one.

And sacrifice must occur on still another level. If we want to break the illusion of separation that keeps us lonely and isolated, we will have to give up the pride and righteousness that keeps us from touching others. We will have to sacrifice our selfishness that demands that all of *our* needs be met before we are willing to give to others. We will give up allowing ourselves to be controlled by the fears that keep us hiding away in our separateness. We will give up controlling situations to suit our own desires, and rather surrender to the needs of the life around us.

Having said all of this, realize that no sacrifice happens before its time. I am not suggesting that people give up any of the things mentioned above until they are ready. The intentions we create will reveal what needs to be given and when. And even though we may give up a great deal, it will never be more than we can afford. It is only through our sacrifice that we find true touch.

HEARTBREAK

I used to be a Prince,
attended by a royal entourage
with all the trappings
of such a position
Now I wander the streets
crying as a fool.
I used to presume our Father
in Heaven bestowed wealth
Now I see Poverty
is His true Gift...[20]

– Lee Lozowick

There comes a time when we turn to touch because we see that we are suffering tremendously from the chasm of separation that keeps us isolated from both ourselves and the ones we love. We intentionally bring physical touch into our lives because we are suffering, and we have admitted it. Because we intuitively know that there is another option to isolation, we grasp, with often frightened and timid hands toward that other option. We get "in touch" because living "out of touch" condemns us to a life of emptiness. We touch others because, having seen our own suffering, their pain no longer eludes us, no matter how clever the guise. Although once propelled into touch by the awareness of a lack within ourselves or by the wish to heal a wound, slowly something else takes over.

There is a time when the wound is no longer the dominating factor in one's life, when the voice of the essential self speaks louder and more often than voices of falseness and confusion, when one has ceased to be a victim of circumstances and instead has taken responsibility for one's life. Still, however, we find that the wound has not disappeared. Instead, it appears to have deepened, though really it has only revealed itself more. The wound stays because it is part of the human condition. The wound remains in order to keep us humble and to remind us of the suffering of those around us.

We even begin to feel gratitude for this wound, for its tender pain, because it has helped us to understand ourselves, and in doing so we have found that we understand others as well. The compassion we feel as a result brings us into genuine relationship with others. This ongoing work of relationship—sharing in our humanness, in our joy, and even in our suffering— becomes the bridge across the very suffering from which it stems. Our pain is what connects us to others, and somehow in the seeming hopelessness of the situation we feel hope because the force of our love has mysteriously grown this within us.

This is the wound of heartbreak.

The way of touch *is* the path of heartbreak and service. The way of touch is what emerges when you see the big picture: the vision of a nation of hungry ghosts...the underlying alarm in people, and the insanity of distraction that has been created in an effort to wield power and to create a fantasy-substitute for reality. You take your first steps on the way of touch when you begin to feel the poison and toxicity of the environment, and to understand the way you have been manipulated into believing it is good, or when the fact of *your* birth and infancy become alive in your mind—what you must have gone through as an infant and young child; what your needs must have been; how your mother and father because of their mother and their father, could not have possibly fulfilled those needs; how your parents' unconsciousness broke your tender heart while they were just doing the best they could. You walk the way of touch when you realize the amount of affection and acknowledgment you needed as a child just because you existed, and not because of your achievements or who you would become for your parents, or when you start to see that your essential self went underground, and that who you think you are is a series of personality traits that you took on in order to make life bearable, and how you deadened yourself to make this all right. You experience the way of touch when you grasp the rage and helplessness that this has left you with—the same helplessness that is in all the people in the society that surrounds you...the people who have suffered as much loss as you have but who have coped with it by wielding power over others, by acquiring, commanding, abusing. You progress on the path of touch when you see what all of this aggression leads to, and you admit that it is as much in you as anybody else, and when you feel the loneliness of the too-large gap between yourself and others—your apprehension of them, your feelings of inferiority, superiority, hatred even. The way of touch is revealed when denial finally makes sense—when you see the necessity of it and can comprehend how it is present in every individual and every institution because reality *is* often too much to bear...when you start to suspect that life isn't how you imagined it...when you grasp that it is truly nobody's fault, that there is nobody to blame...when you recognize that you are paying for this false sense of security with your life. You are living the way of touch when what once seemed to be solid ground becomes shifting sand...when you admit how much of your life is a lie...when you see the pain of the lie in other people's eyes.

Where does this path eventually lead? To heartbreak. Why then bother to walk it? Because with heartbreak comes a clear vision of reality.

Your heart breaks—not as it did when you were an infant, but instead breaks open. You want it to break open. Yet even when you have been wholly broken you remain right where you are, vulnerable and raw, realizing that all you want to do is help others with the suffering in their lives. You want to do everything you can to bring more love to them. You want to hold crying children, and make babies feel welcome and beautiful and important. You want to touch people with your presence and your concern. You want to be kind and to share your resources. You want to see the beauty in others and help them to see it for themselves. You want the food that you place into your body and that you serve to others to be prepared with love. You want to treat everything—animate and inanimate—with care and respect. You want to use every moment to do whatever is needed to help heal, to bring joy, to celebrate the beauty in the small things of life. You want to do no harm. You want to live in gratitude for your heartbreak because it has made you real. This is the life of service—the way of touch.

N O T E S

Chapter One

[1] Montagu, Ashley, *Touching: The Human Significance of the Skin*—Second Edition (New York: Harper and Row, 1978), pp. 243-244.

[2] Holdhagen, Daniel, *Hilter's Willing Executioners: Ordinary Germans and the Holocaust* (New York: Knopf, 1996).

[3] Nouwen, Henri J.M., *The Wounded Healer* (New York: Doubleday, 1972), pp. 5-7.

[4] Note: if you are in a 12-step program or have any intention of engaging in one, I *do not* recommend these 12 steps!

[5] Morris, Desmond, *Intimate Behavior* (New York: Random House, 1971), p. 173.

[6] Freedman, David, "Playing God: The Making of Artificial Life," *Discover*, August, 1992.

[7] On the physical level, electromagnetic radiation [EMR] has been linked to leukemia and brain cancer in children, according to a report released by the Environmental Protection Agency. It is not only the computers and high power office equipment that are releasing this "electro-pollution," but televisions, microwaves, X-rays, the high transmission lines that cross the nation, and nuclear power lines. The average person is now

exposed daily to 100 million times the electromagnetic radiation than that which his grandparents were. EMR is also suspected to be linked with headaches, memory loss, rare tumors, poor crops, birth defects, and depression. Referenced in: *Advanced Living Technologies,* 2442 Meade St., Denver, CO 88021-4440.

[8] Ibid.

[9] Edelman, Vladimir, "Touch Goes High-Tech," *Psychology Today*, January/February, 1996.

[10] Ibid.

[11] Levy, Steven, "No Place for Kids? A Parents' Guide to Sex on the Net," *Newsweek,* July 3, 1995.

[12] Ibid.

[13] Kozol, Jonathan, *Amazing Grace* (New York: Crown Publishers Inc., 1995), p. 64.

[14] Ibid, p. 31.

[15] Ibid, p. 223.

[16] Ibid, p. 188.

[17] liars, gods and beggars, "Can You Trust?," performed on *Vuja Dé,* copyright by Lee Lozowick and Stan Hitson.

[18] Putnam, Robert D., "Bowling Alone," *Current,* June, 1995.

[19] Rilke, Rainer Maria, *"Fragment of an Elegy,"* Mitchell, Stephen (editor). *Selected Poetry of Rainer Maria Rilke* (New York: Vintage Books/Random House, 1994), p.215.

[20] Editors, *Tawagoto,* Winter, 1988.

[21] Montagu, p. 315.

[22] Pearce, Joseph Chilton, *Evolution's End* (San Francisco, CA: HarperSan Francisco, 1992), p. 197.

Chapter Two

[1] Leboyer, Frederick, *Loving Hands* (New York: Alfred A. Knopf, 1976), p.16.

[2] Chang, Jolan, *The Tao of Love and Sex: The Ancient Chinese Way to Ecstasy* (New York: E.P. Dutton, 1977), p. 27.

[3] Liedloff, Jean, *The Continuum Concept* (Reading, MA: Addison-Wesley Publishing Co., 1977), p. 48.

[4] Rilke, Rainer Maria, "Fourth Elegy," *The Selected Poetry of Rainer Maria Rilke* (New York: Vintage International, 1980), p. 173.

[5] Watson, John B., *Psychological Care of Infant and Child* (New York: Arno Press, 1972), p. 81.

[6] Montagu, Ashley, *Touching: The Human Significance of the Skin—Second Edition* (New York: Harper and Row, 1978), p. 142.

[7] Pearce, Joseph Chilton, *Magical Child* (New York: Bantam, 1977), p. 72.

[8] Pearce, Joseph Chilton, *Magical Child Matures* (New York: E.P. Dutton, 1985), p. 25.

[9] Ibid., p.34.

[10] Dunham, Carroll, et. al., *Mamatoto* (New York: Penguin Books, 1991), p.4.

[11] Montagu, p.142.

[12] Pearce, *Magical Child*, p. 72.

[13] Stettbacher, J. Konrad, *Making Sense of Suffering* (New York: Dutton, 1991), p. 112.

[14] Klein, Paul F., "The Needs of Children," *Mothering*, Spring, 1995.

[15] Editorial staff, *Hohm Sahaj Mandir Study Manual, Volume I* (Prescott, AZ: Hohm Press, 1996), p. 79.

[16] Prescott, James, "Failure of Pleasure As a Cause of Drug/Alcohol Abuse and Addictions," *The Truth Seeker*, September/October 1989.

[17] Pearce, *Magical Child*, p. 61.

[18] Montagu, p. 176.

[19] Prescott, James, "Alienation of Affection," *Psychology Today,* December, 1979.
[20] Prescott, James, "Somatosensory Affectional Deprivation (SAD) Theory of drug and alcohol use," in *Theories On Drug Abuse: Selected Contemporary Perspectives*, 1980. Dan J. Lettieri, Mollie Sayers and Helen Wallenstien Pearson, Eds.) NIDA Research Monograph, March 30, 1980. National Institute on Drug Abuse, Department of Health and Human Services. Rockville, MD).
[21] Prescott, James, Time Life Documentary, "Rock a Bye Baby," premiered at the 1970 White House Conference On Children.
[22] Pearce, Joseph Chilton, "Interview," *Mothering,* Spring, 1985.
[23] Verny, Thomas, M.D. and John Kelly, *The Secret Life of the Unborn Child* (New York: Dell, 1981), pp.12-13.
[24] Verny, Thomas, M.D. and Pamela Weintraub, *Nurturing the Unborn Child* (New York: Delcorte Press, Bantam Doubleday Dell Publishing Group, 1991), p.xxiii.
[25] Ibid., p. 22-23.
[26] Ibid., p. 20-21.
[27] Sears, William, *The Fussy Baby* (Franklin Park, IL: La Leche League International, 1985), p. 29.
[28] Sears, p. 30.
[29] Sears, William, *Keys to Becoming a Father* (New York: Barrons, 1991), p. 3.
[30] Leboyer, Frederick, *Birth Without Violence* (New York: Alfred A. Knopf, 1984), pp. 59-60.
[31] Davis-Floyd, Robbie, "Hospital Birth as a Technocratic Right of Passage," *Mothering,* Summer, 1993.
[32] Verny, p.xxvi.
[33] Kitzinger, Sheila, *Homebirth: The Essential Guide to Giving Birth Outside of the Hospital, 1st ed.* (New York: Dorling Kindersley, 1991), pp. 43-44.

[34] Buhler, Karen, M.D., *Raising Issues.* Summer, 1997.
[35] Department of Health, "Changing Childbirth Report of the Expert Maternity Group." London, England: HMSO, 1993.
[36] Ashford, Janet Issacs, "The History of Midwifery in the United States," *Mothering,* Winter, 1990.
[37] Ibid.
[38] There was a worldwide boycott on Nestle baby formula when it became publicly recognized that infant mortality rates soared, particularly in the Third World, when infants were given baby formula over breast milk. (Palmer, Gabrielle. "The Politics of Infant Feeding," *Mothering,* Summer, 1991.)
[39] "Letter to the Editor," *Mothering,* Spring, 1994, p.15.
[40] Montagu, p. 264.
[41] Pearce, *Magical Child Matures,* p. 31.
[42] Excerpted from Glick, Daniel, "Rooting for Intelligence," *Newsweek,* Spring/Summer 97 Special Edition, p.32.
[43] Excerpted from "Breast Feeding and Intelligence," *Pediatrics for Parents,* July/August, 1993, p.12.
[44] Leidloff, pp. 36-37.
[45] Ibid, p. 37.
[46] Dunham, pp. 164-165.
[47] Prescott, James, "The Origins of Human Love and Violence," p. 8.
[48] Rilke, "The Third Elegy," p. 163
[49] Sears, William, *SIDS: A Parent's Guide to Understanding and Preventing Sudden Infant Death Syndrome* (New York: Little, Brown and Company, 1995), p. 102.
[50] Excerpted from Gordon, David, "Preventing a Hard Day's Night," *Newsweek, Special Edition,* Spring/Summer 1997, p.56.
[51] Ibid, p. 125.
[52] Ibid, pp. 102-103.

[53] Field, Tiffany, "Massage Therapy for Infants and Children." *Developmental and Behavioral Pediatrics,* Vol. 16, No. 2., April, 1995.p. 106.

[54] Ibid., Research attributed to Auckett, A.D., *Baby Massage, NY*: Newmarket Press, 1981; McClure, VS, *Infant Massage,* NY: Bantam, 1989.

Chapter Three

[1] Kreyche, Gerald F., "Day Care—The New Surrogacy," *USA Today,* September, 1989.

Chapter Four

[1] Leboyer, Frederick, *Birth Without Violence* (New York: Alfred A. Knopf, 1984), pp. 110-111.

[2] Prescott, James. "Hollywood Violence .. First Do No Harm," *Imagine a World of Wanted Children,* August, 1995.

[3] In Sweden in July, 1979, The Swedish Parenthood and Guardianship Code created the clause: "The child may not be subjected to physical punishment or other injurious or humiliating treatment." The first of its kind in national legislation, it inspired similar laws to be passed in Norway, Finland, Denmark and Austria. (Haeuser, Adrienne Ahlgren, "Swedish Parents Don't Spank," *Mothering,* Spring, 1992.)

[4] Hyde, Margaret O., *Sexual Abuse—Let's Talk About It* (Philadelphia, PA: The Westminster Press, 1984), p.15.

[5] Lozowick, Lee, (cited in) *Hohm Sahaj Mandir Study Manual, Vol. 2* (Prescott, AZ: Hohm Press, 1996) , p. 132.

[6] In the United States, 29 states still allow corporal punishment in their schools. Almost every European country has outlawed this practice beginning in the 1800s. (Fathman, Robert E., *Mothering,* Winter, 1991.)

[7] *NEA Today,* February, 1994.

[8] Ibid.

[9] Seligman, Katherine, "Sex Laws for Schools," *San Francisco Examiner,* May 30, 1993.

[10] Ibid.

[11] Ibid.

[12] Leboyer, Frederick, *The Art of Breathing* (Paris, France: Editions de Seuil), p. 135.

[13] The majority of blatant abuse is perpetrated by fathers and men, but women are accountable for many of the subtler, yet equally damaging forms of abuse including passive witnessing, neglect, manipulation, failure to tend to the infants needs, etc.

[14] For further study in this area, see: Bradshaw, John, *Family Secrets* (New York: Bantam, 1995); and Sheldrake, Rupert, *The Presence of the Past* (New York: Random House, 1988).

[15] Gorney, Roderic, "Hope for Humankind," *The Humanist,* January/February,1996.

[16] Prescott, James, *NIH Violence Research Initiatives*, p. 8-9; *The Origins of Human Love and Violence,* p. 4.

[17] Ibid.,"The Origins of Human Love and Violence, p.7.

[18] Ibid., p. 15.

[19] Kozol, Jonathan, *Amazing Grace* (New York: Crown Publishers Inc.), p. 239.

[20] The edited version is found in *The New York Times,* August 21, 1996; the unedited version can be found in Issue 38 of *Wellness Associates Journal,* 123 Wildwood Trail, Afton, VA, 22920.

[21] Montagu, Ashley, *Touching: The Human Significance of the Skin*—Second Edition (New York, Harper and Row, 1978), p. 206.

[22] Ospensky, P.D., (Quoting G.I. Gurdjeiff), *In Search of the Miraculous* (New York: Harcourt Brace Jovanovich Inc., 1949), p. 54.

[23] Reprinted with permission from: Lenox, Annie and David Stewart. "This City Never Sleeps." D'N'A' Ltd., BMG Music Publishing, Ltd., 1983.

[24] Campbell, Greg, Interview in Germany, June, 1996.

[25] It is common in dysfunctional families for a child to "take on" the symptoms of the parents' conflicts. In family therapy, the child who does this is referred to as the "identified patient," or the "scapegoat," as the finger of blame is pointed at the child so that the parents can avoid facing their own problems. A detailed explanation of this phenomenon can be found in Napier, Augustus and Carl Whittaker, *The Family Crucible* (New York: Harper and Row, 1978).

[26] Gorney, Roderic, "Hope for Humankind," *The Humanist,* January/February, 1996.

[27] To do any justice to the complexity, intricacies, and consequences of this proposition would require a full volume, at least. For the purposes of understanding the consequences of the lack of touch in our lives, however, an overview is included here.

[28] Liedloff, Jean, *Continuum Concept* (Reading, MA: Addison-Wesley Publishing Co., 1977), p. 143.

[29] Montagu, pp. 243-244.

[30] Rilke, Rainer Maria, "The Third Elegy," *The Selected Poetry of Rainer Maria Rilke* (New York: Vintage International, 1980), p.165.

[31] My first book, *When Sons and Daughters Choose Alternative Lifestyles* (Hohm Press, 1996), was written in an attempt to bridge this gap of alienation and separation that often occurs in families who are faced with this situation. The misunderstandings are largely due to

cultural conditioning and social stereotypes and misconceptions.

[32] Lozowick, Lee, Public Talk, May, 1996.

[33] Stettbacher, J. Konrad, *Making Sense of Suffering* (New York: Dutton, 1991), p. 11.

Chapter Five

[1] Rilke, Rainer Maria, "Turning-point," *The Selected Poetry of Rainer Maria Rilke* (New York: Vintage Books/ Random House, 1984), p. 135.

[2] Referenced in: *Advanced Living Technologies,* 2442 Meade St., Denver, CO 88021-4440.

[3] Crenshaw, Theresa L. *The Alchemy of Love and Lust.* New York: Pocket Books (Simon & Schuster), 1996.

[4] It is important to be able to distinguish between a skilled therapist and an unskilled one. The chapter entitled "Sources of Help" in my first book, *When Sons and Daughters Choose Alternative Lifestyles* (Hohm Press, 1996), details the process of selecting a qualified therapist.

[5] Field, Tiffany, "Massage Therapy for Infants and Children," *Developmental and Behavioral Pediatrics,* Vol. 16, No. 2, April, 1995.

[6] Sunshine, W., Filed, T.M. et al. "Chronic fatigue syndrome: Massage therapy effects on depression and fatigue." *Journal of Chronic Fatigue Syndrome,* 4, 1997, pp. 43-51.

[7] Field, T., Delamanter, A.M. et al. "Diabetic children's adherence and glucose levels improve after touch therapy," *Diabetes Spectrum,* 1997; Field T., Seligman, S. et al. "Alleviating post-traumatic stress in children following Hurricane Andrew," *Journal of Applied Developmental Psychology,* 17, 1996, pp.37-50.

[8] Field, T., Morrow, C. et al. "Massage therapy reduces anxiety in child and adolescent psychiatric patients." *Journal of the American Academy of Child and Adolescent Psychiatry,* 31 (1), 1992, pp. 125-131.

[9] Ironson, Gail and Tiffany Field et. al., "Massage Therapy is Associated with Enhancement of the Immune System's Cytotoxic Capacity." *Intern. J. Neuroscience,* 1995.

[10] Krieger, Dolores, "Therapeutic Touch: The Imprimatur of Nursing," *American Journal of Nursing,* 75 (1975), 784-87.

Chapter Six

[1] Montagu, Ashley, *Touching: The Human Significance of the Skin*—Second Edition (New York: Harper and Row, 1978), p. 166.

[2] Lozowick, Lee, *The Alchemy of Transformation* (Prescott, AZ: Hohm Press, 1996), p.70.

[3] Cloud, John, "Ivy League Gomorrah?," *Time,* September 22, 1997.

[4] Leidloff, Jean, *The Continuum Concept* (Reading, MA: Addison-Wesley Publishing Co., 1977), p. 154.

[5] Rilke, Rainer Maria, " The Third Elegy," *The Selected Poetry of Rainer Maria Rilke,* (New York: Vintage International, 1980), p. 167.

[6] Lozowick, Lee, *The Alchemy of Love and Sex* (Prescott, AZ: Hohm Press, 1996), p. 209.

[7] Lozowick, Lee, *Living God Blues* (Prescott, AZ: Hohm Press, 1984), p.22.

[8] Emma, Ushanda io, "Life With the Pygmies, *Mothering,* Summer 1988.

[9] Cummings, E.E., "since feeling is first," (poem copyright 1926); from: Kennedy, Richard (editor), *E.E. Cummings Selected Poems* (New York: Liveright, 1994), p. 99.

Chapter Seven

[1] Cummings, E.E., "somewhere I have never travelled,gladly beyond," (poem copyright 1931); from: *E.E. Cummings Complete Poems 1904-1962* (New York: Liveright, 1973) p. 367.

[2] Pearce, Joseph Chilton, (Interview), *Mothering,* Spring 1985.

[3] It is for this reason that scientific discoveries are often made simultaneously in different parts of the world—the objective knowledge in the universe suddenly coming into a configuration that is accessible to the open receptacle of the scientific mind. By the time our mind is consciously aware of a piece of information, that information has already been "alive" for eons.

[4] Dossey, Larry, *Prayer is Good Medicine* (CA: HarperSan Francisco), pp. 3-4

[5] Byrd, Randolph C., "Positive Therapeutic Effects of Intecessory Prayer in a Coronary Care Unity Population," *Couthern Medical Journal* 81, no. 7 (July, 1988): 826-29.

[6] Jigme Rinpoche, Public Talk, December 1995, Adelaide, Australia.

[7] Tweedie, Irina, *Chasm of Fire* (Wiltshire, England: Element Books, 1979), p. 202.

[8] Stevens, John, *The Essence of Aikido: Spiritual Teachings of Morihei Ueshiba* (New York: Kodansh America Inc., 1993), p. 13.

[9] Gold, E.J., *The Joy of Sacrifice* (Prescott, AZ: IDHHB Inc. and Hohm Press, 1978), p. 106.

[10] Nouwen, Henri, *The Wounded Healer* (New York: Doubleday, 1972), p. 91.

[11] Mother Teresa of Calcutta, *Life in the Spirit* (San Francisco, CA: Harper and Row, 1983), p. 93.

[12] Conway, Timothy, *Women of Power and Grace* (Santa Barbara, CA:The Wake Up Press, 1984), p. 264.

[13] Ibid., p. 112.

[14] Sister Chan Khong, *Learning True Love* (Berkeley, CA: Parallax Press, 1993), p. 249.

[15] Conari Press (editors), *Random Acts of Kindness* (Berkeley, CA: Conari Press, 1993), p. 37.

[16] Conari Press (editors), *Kids' Random Acts of Kindness* (Berkeley, CA: Conari Press, 1994), p. 105-6.

[17] Wilber, Ken, *Grace and Grit* (Boston, MA: Shambhala Publications, 1991), p. 405.

[18] Miller, Ronald S., *As Above, So Below* (Los Angeles, CA: J.P. Tarcher, 1992), p. 283.

[19] Gold, E.J., p. 28.

[20] Lozowick, Lee, *Poems of a Broken Heart* (Madras, India: Sister Nivedita Academy, 1993), p. 23

BIBLIOGRAPHY AND RECOMMENDED READING

Books

Ainsworth, Mary. *Patterns of Attachment.* Hillsdale, NJ: Lawrence Earlbaum Associates, 1978.

Anderson, Sherry Ruth and Patricia Hopkins. *The Feminine Face of God.* New York: Bantam, 1992.

Attwood, Charles. *Dr. Attwood's Low-Fat Prescription for Kids.* New York: Viking, 1995.

Barker, Sarah. *The Alexander Technique.* New York: Bantam Books, 1991.

Barnard, Kathryn E. and T. Berry Brazelton. *Touch: The Foundation of Experience.* Madison, CT: International Universities Press, 1990.

Berends, Polly Berrien. *Whole Child/Whole Parent.* New York: Harper & Row, 1983.

Bowlby, John. *A Secure Base.* New York: Basic Books, 1988.

Bradshaw, John. *Family Secrets.* New York: Bantam, 1995.

Brown, Barbara. *New Mind, New Body.* New York: Harper and Row, 1974.

Cambell, Gregory. *I Put It In the Nursing Notes: About this Radiance in Our Eyes!* Prescott, AZ: Hohm Press, 1987.

Caplan, Mariana. *When Holidays are Hell...! A Guide to Surviving Family Gatherings.* Prescott, AZ: Hohm Press, 1997.

—. *When Sons and Daughters Choose Alternative Lifestyles.* Prescott, AZ: Hohm Press, 1996.

Chamberlain, David. *Babies Remember Birth.* New York: Ballantine Books, 1988.

Chang, Jolan. *The Tao of Love and Sex: The Ancient Chinese Way to Ecstasy.* New York: E.P. Dutton, 1977.

Chan Khong, (Sister). *Learning True Love,* Berkeley, CA: Parallax Press, 1993.

Colton, Helen. *The Gift of Touch: How Physical Contact Improves Communication, Pleasure, and Health.* New York: Seaview and Putnam, 1983.

Conari Press (editors). *Random Acts of Kindness.* Berkeley, CA: Conari Press, 1993.

—. *Kids' Random Acts of Kindness.* Berkeley, CA: Conari Press, 1994.

Conway, Timothy. *Women of Power and Grace.* Santa Barbara, CA: The Wake Up Press, 1994.

Crenshaw, Theresa L. *The Alchemy of Love and Lust.* New York: Pocket Books (Simon & Schuster), 1996.

Dossey, Larry. *Prayer is Good Medicine.* CA: HarperSan Francisco, 1996.

Dougans, Inge, and Suzanne Ellis. *The Art of Reflexology.* Rockport, MA: Element Books, 1992.

Downing, George. *The Massage Book.* New York: Random House, 1972.

Dunham, Carroll, et. Al. *Mamatoto.* New York: Penguin Books, 1991.

Editorial board. *Massage: Total Relaxation.* Alexandria, Virginia: Time-Life Books, 1987.

Editorial staff. *Hohm Sahaj Mandir Study Manual (Volumes I and II).* Prescott, AZ: Hohm Press, 1996.

Feurstein, Georg, and Stephan Bodian. *Living Yoga.* New York: Jeremy P. Tarcher, 1993.

Field, Reshad. *Here to Heal.* Longmead, Shaftesbury, Dorset: Elemental Books, 1985.

Fields, Rick. *Chop Wood, Carry Water.* Los Angeles: J.P. Tarcher, distributed by Houghton Mifflin, 1984.

Fifer, Steve, and Sharon Sloan Fiffer. *50 Ways to Help Your Community.* New York: Doubleday, 1994.

Ford, Clyde W. *Compassionate Touch.* New York: Simon and Schuster, 1993.

Gaskin, Ina May. *Spiritual Midwifery.* Summertown, TN: The Book Publishing Co., 1978.

Gawain, Shakti. *Path of Transformation: How Healing Ourselves Can Change the World.* Mill Valley, CA: Nataraj, 1993.

Gold, E.J. *The Joy of Sacrifice.* Prescott, Arizona: IDHHB Inc. and Hohm Press, 1978.

Goldhagen, Daniel Jonah. *Hitler's Willing Executioners.* New York: Alfred A. Knopf, 1996.

Grigg, Ray. *The Tao of Relationships.* New York: Bantam, 1988.

Hendin, Herbert. *The Age of Sensation.* New York: Norton, 1975.

Henley, Nancy M. *Body Politics: Power, Sex and Nonverbal Communication.* Englewood Cliffs, NJ: Prentice-Hall, 1977.

Hyde, Margaret O. *Sexual Abuse—Let's Talk About It.* Philadelphia, PA: The Westminster Press, 1984.

Johnson, Don. *The Protean Body: A Rolfer's View of Human Flexibility.* New York: Harper Colophon Books, 1977.

Kitzinger, Sheila. *Homebirth: The Essential Guide to Giving Birth Outside the Hospital.* New York: Dorling Kindersley, 1991.

—. *The Complete Book of Pregnancy and Childbirth.* New York: Alfred A. Knopf, 1985.

Kozol, Jonathan. *Amazing Grace.* New York: Crown Publishers Inc., 1995.

Krieger, Dolores. *Living the Therapeutic Touch.* New York: Dodd, Mead & Company, 1987.

—. *Accepting Your Power to Heal: The Personal Practice of Therapeutic Touch.* Santa Fe, NM: Bear&Co., 1993.

—. *Therapeutic Touch: How to Use Your Hands to Help or to Heal.* Englewood Cliffs, NJ: Prentice-Hall, 1979.

—. *Therapeutic Touch Workbook: Ventures in Transpersonal Healing.* Santa Fe, NM: Bear&Co., 1997.

Leboyer, Frederick. *Birth Without Violence.* New York: Alfred A. Knopf, 1984.

—. *Inner Beauty, Inner Light.* New York: Alfred A. Knopf, 1978.

—. *Loving Hands.* New York: Alfred A. Knopf, 1976.

—. *The Art of Breathing.* Paris, France: Editions de Seuil, 1985.

Levine, Stephen. *Who Dies?* Garden City, New York: Anchor Press-Doubleday, 1982.

—. *A Year to Live: How to Live This Year as if It Were Your Last.* New York: Bell Tower, 1997.

Liedloff, Jean. *The Continuum Concept.* Reading, MA: Addison-Wesley Publishing Co., 1977.

Lozowick, Lee. *Conscious Parenting.* Prescott, AZ: Hohm Press, 1997.

—. *Living God Blues.* Prescott, AZ: Hohm Press, 1984.

—. *The Alchemy of Love and Sex.* Prescott, AZ: Hohm Press, 1996.

—. *The Alchemy of Transformation.* Prescott, AZ: Hohm Press, 1996.

—. *Poems of a Broken Heart.* Madras, India: Sister Nivedita Academy, 1993.

Macrae, Janet. *Therapeutic Touch: A Practical Guide.* New York: Alfred A. Knopf, Inc., 1987.

Macy, Joanna. *Despair and Personal Power in the Nuclear Age.* Philadelphia, PA: New Society Publishers, 1983.

Mander, Henry. *Four Arguments for the Elimination of Television.* New York: Quill, 1977.

Mansfield, Lynda Gianforte and Christopher H. Waldmann. *Don't Touch My Heart.* Colorado Springs, CO: Piñon Press, 1994.

Martin, Chia. *The Art of Touch: A Massage Manual for Young People.* Prescott, AZ: Hohm Press, 1996.

Miller, Alice. *Banished Knowledge: Facing Childhood Injuries.* New York: Doubleday, 1990.

—. *For Your Own Good: Hidden Cruelty in Child-Rearing and the Roots of Violence.* New York: Farrar, Straus, Giroux, 1984.

—. *Thou Shalt Not Be Aware: Society's Betrayal of the Child.* New York: Meridian Books (Penguin), 1990.

Miller, Ronald S. *As Above, So Below.* Los Angeles, CA: J.P. Tarcher, 1992.

Montagu, Ashley. *Touching: The Human Significance of the Skin,* (Second Edition). New York: Harper and Row, 1978

—. *Growing Young,* (Second Edition). Granby, MA: Bergen & Garvey. 1989.

—. *On Being Human,* (Second Edition). New York: Hawthorn Books, 1967.

—. *The Nature of Human Aggression.* New York: Oxford University Press, 1976.

—. *Touching: The Human Significance of the Skin,* (Second Edition). New York: Harper and Row, 1978.

—. *The Dehumanization of Man.* New York: McGraw-Hill, 1983.

Mother Teresa of Calcutta. *Life in the Spirit.* San Francisco, CA: Harper and Row, 1983.

—. *The Love of Christ: Spiritual Counsels.* San Francisco, CA: Harper and Row, 1982.

Morris, Desmond. *Intimate Behavior.* New York: Random House, 1971.

Northrup, Christiane, M.D. *Women's Bodies, Women's Wisdom: Creating Physical and Emotional Health and Healing.* New York: Bantam Books, 1994.

Nouwen, Henri J.M. *The Wounded Healer.* New York: Doubleday, 1972.

Otto, Herbert A. *Love Today: A New Exploration.* New York: Association Press, 1972.

Ouspensky, P.D. *In Search of the Miraculous.* New York: Harcourt Brace Jovanovich Inc., 1949.

Pearce, Joseph Chilton. *Evolution's End.* San Francisco, CA: HarperSanFrancisco, 1992.

—. *Magical Child.* New York: Bantam, 1977.

—. *Magical Child Matures.* New York: E.P. Dutton, 1985.

Peck, M. Scott. *People of the Lie.* New York: Simon and Schuster, 1983.

Piercy, Marge. *The Moon is Always Female.* New York: Alfred A. Knopf, 1988.

Prescott, James. From: *Violent Behavior: Assessment and Intervention.* Great Neck, New York: PMA Publishing Corp., 1990.

Rahula, Walpola. *What the Buddha Taught.* New York: Grove Press Inc., 1959.

Reich, Willhelm. *The Function of the Orgasm.* New York: Farrar Straus and Giroux, 1973.

Rilke, Rainer Maria. *Selected Poems of Rainer Maria Rilke.* New York: Harper and Row, 1981.

—. *The Selected Poetry of Rainer Maria Rilke.* Edited by Stephen Mitchell. New York: Vintage International, 1980.

Schaef, Anne Wilson. *When Society Becomes an Addict.* San Francisco, CA: Harper and Row, 1987.

Sears, William, M.D. and Martha Sears. *The Baby Book: Everything You Need to Know About Your Baby—From Birth to Age Two.* New York: Little, Brown and Company, 1993.

Sears, William, M.D. *Keys to Becoming a Father.* New York: Barrons, 1991.

—. *SIDS: A Parent's Guide to Understanding and Preventing Sudden Infant Death Syndrome.* New York: Little, Brown and Company, 1995.

—. *The Fussy Baby.* Franklin Park, IL: La Leche League International, 1985.

Sellner, Edward C. *Wisdom of the Celtic Saints.* Notre Dame, Indiana: Ave Maria Press, 1993.

Sen, Soshitsu. *Tea Life, Tea Mind.* New York: Weatherhill, 1979.

Sheldrake, Rupert. *The Presence of the Past.* New York: Random House, 1988.

Siegel, Bernie, M.D. *Love, Medicine and Miracles.* New York: Harper and Row, 1986.

Simon, Sidney B. *Caring, Feeling, Touching.* Niles, IL: Argues Communications, 1976.

Shiva, Shahram T. *Rending the Veil: Literal and Poetic Translations of Rumi.* Prescott, AZ: Hohm Press, 1995.

Somé, Malidoma. *Of Water and the Spirit: Ritual, Magic, and Initiation in the Life of an African Shaman.* New York: Putnam, 1994.

Sorell, W. *Story of the Human Hand.* London: Weidenfeld & Nicholson, 1968.

Spink, Kathryn. *Gandhi.* London, England: Hamish Hamilton Children's Books, 1984.

Stettbacher, J. Konrad. *Making Sense of Suffering.* New York: Dutton, 1991.

Stevens, John. *The Essence of Aikido: Spiritual Teachings of Morihei Ueshiba.* New York: Kodansh America Inc., 1993.

Stransky, Judith and Robert B. Stone. *The Alexander Technique.* New York: Beaufort Books, 1981.

Thie, John F. and Mary Marks. *Touch for Health.* Pasadena, CA: Touch for Health Foundation, 1973.

The Troops for Truddi Chase. *When Rabbit Howls.* New York: E.P. Dutton, 1987.

Thevenin, Tine. *The Family Bed.* Wayne, New Jersey: Avery Publishing Group, 1987.

—. *Mothering and Fathering: The Gender Differences in Child Raising.* Wayne, NJ: Avery Publishing Group, 1993.
Tweedie, Irina. *The Chasm of Fire.* Wiltshire, England: Element Books, 1979.

Upton, Charles. *Doorkeeper of the Heart: Versions of Rabi'a.* Putney, VT: Threshold Books, 1988.

Verny, Thomas, M.D. and John Kelly. *The Secret Life of the Unborn Child.* New York: Dell, 1981.

—. *Nurturing the Unborn Child.* New York: Dell, 1992.

Watson, John B. *Psychological Care of Infant and Child.* New York: Arno Press, 1972.

Wilber, Ken. *Grace and Grit.* Boston, MA: Shambhala Publications, 1991.

Zilbergeld, Bernie. *The New Male Sexuality.* New York: Bantam Books, 1992.

Articles

Ashford, Janet Issacs. "The History of Midwifery in the United States." *Mothering,* Winter, 1990.

Associated Press. "An Electrifying New Hazard." *U.S. News & World Report,* March 30, 1987.

Associated Press. "Report linking electric fields, cancer delayed by White House." *Arizona Republic,* December 14, 1990.

Baker, Beth. "The Changing Face of Social Work." *Common Boundary,* January/February, 1994.

Bar-Yam, Naomi. "The Nestle Boycott." *Mothering,* Winter, 1995.

Biemiller, Lawrence. "Lessons From a Sobering Age: 'Touch People. Hug Them When You Need To.'" *The Chronicle of Higher Education,* October, 1994.

Blackman, Nancy B. "Pleasure and Touching: Their Significance in the Development of the Preschool Child." Paper presented at the International Symposium on Childhood and Sexuality, Montreal, Sept. 1979.

Bassoff, Evelyn Silten. "Healing Love." *Mothering,* Spring, 1991.

Brandeis-McGunigle, Gayle. "Childhood and Ritual in Bali." *Mothering,* Fall, 1991.

Burton, Arthur and Louis G. Heller. "The Touching of the Body." *Psychoanalytic Review,* Vol. 51, no. 1, Spring, 1964.

Byrd, Randolph C., "Positive Therapeutic Effects of Intercessory Prayer in a Coronary Care Unity Population." *Southern Medical Journal,* 81, no. 7, July, 1988.

Clay, Vidal S. "The Effect of Culture on Mother-Child Tactile Communication." *Family Coordinator,* Vol. 17, 1968.

Davis-Floyd. "Hospital Birth As a Technocratic Rite of Passage." *Mothering,* Summer, 1993.

Edelman, Vladimir. "Touch Goes High-Tech: Dispatch from Virtual Reality." *Psychology Today,* January/February, 1996.

Editor. "Debate: Should You Ever Touch A Student," *NEA Today,* February, 1994.

Emma, Ushanda io. "Life With The Pygmies." *Mothering,* Summer, 1988.

Falbel, Aaron. "The Computer as a Convivial Tool." *Mothering,* Fall, 1990.

Fathman, Robert E. "Child Abuse." *Mothering,* Winter, 1991.

Ferguson, Eugene S. "How Engineers Lose Touch." *American Heritage of Invention & Technology,* Winter, 1993.

Field, Tiffany. "Massage Therapy for Infants and Children." *Developmental and Behavioral Pediatrics,* Vol. 16, No. 2., April, 1995.

Field, T., Delamanter, A.M. et al. "Diabetic children's adherence and glucose levels improve after touch therapy," *Diabetes Spectrum,* 1997

Field T., Seligman, S. et al. "Alleviating post-traumatic stress in children following Hurricane Andrew," *Journal of Applied Developmental Psychology,* 17, 1996, pp.37-50.

Field, T., Morrow, C. et al. "Massage therapy reduces anxiety in child and adolescent psychiatric patients." *Journal of the American Academy of Child and Adolescent Psychiatry,* 31 (1), 1992, pp. 125-131.

Frankel, Max. "Do Computers Eat Our Paychecks?" *New York Times Magazine,* March, 1996.

Freedman, David H. "Playing God: The Making of Artificial Life." *Discover,* August, 1992.

Gibson, James J. "Observations on Active Touch." *Psychological Review,* Vol. 69, no. 6, November, 1962.

Goldman, Caren. "Prisons can be more than holding tanks. They can be sacred spaces where troubled souls are guided to mature manhood." *Common Boundary,* January/February, 1993.

Gorney, Roderic, "Hope for Humankind," *The Humanist,* January/February, 1996.

Haeuser, Adrienne Ahlgren. "Swedish Parents Don't Spank." *Mothering,* Spring, 1992.

Hollender, Marc. H. and Alexander J. Mewrcer. "The Wish to be Held and to Hold in Men and Women." *Archives of General Psychiatry,* Vol. 33, January, 1976.

Hovey, Pauline O. "The Quality of Mercy." *Common Boundary,* March/April, 1993.

Ironson, Gail and Tiffany Field et. al. "Massage Therapy is Associated with Enhancement of the Immune System's Cytotoxic Capacity." *Intern. J. Neuroscience,* 1995.

Jeffers, Frank E. "Love and the Relatedness of Things." *The Humanist,* January/February, 1996.

Klein, Paul. "The Needs of Children." *Mothering,* Spring, 1995.

Kreyche, Gerald F. "Day Care—The New Surrogacy." *USA Today,* Sept., 1989.

LaCerva, Victor. "Dearest Papa." *Mothering,* Fall, 1984.

Lattin, Don. "The Problems and the Potential of 'Reaching Out.'" *Common Boundary,* September/October, 1991.

Liedloff, Jean. "Normal Neurotics Like Us." *Mothering,* Fall, 1991.

Levy, Steven. "No Place for Kids? A Parents' Guide to Sex on the Net." *Newsweek,* July 3, 1995.

Major, Brenda and Richard Heslin, "Perceptions of Same-Sex and Cross-Sex Touching." Paper presented at Midwestern Psychological Association Conference, Chicago, 1978.

Mathison, Linda. "Birth Memories: Does Your Child Remember?" *Mothering,* Fall, 1981.

Mc Carthy, Anna. "Reach Out and Touch Someone: Technology and Sexuality in Broadcast Ads for Phone Sex." *The Velvet Light Trap,* No. 32, Fall, 1993.

Palmer, Gabrielle. "The Politics of Infant Feeding." *Mothering,* Summer, 1991.

Pearce, Joseph Chilton. (Interview). *Mothering,* Spring, 1985.

Plaut, Michael S. and Heidi B. Ginter. "Sexual Boundaries Between Health Professionals and Clients: A Blueprint for Education." SIECUS Report, June/July, 1995.

Prescott, James. "Alienation of Affection," *Psychology Today,* December, 1979.

—. "Body Pleasure and the Origins of Violence." *The Futurist,* April, 1975.

—. "Failure of Pleasure as a Cause of Drug/Alcohol Abuse and Addictions." *The Truth Seeker,* September/October, 1989.

—. "Hollywood Violence...First Do No Harm," *Imagine a World of Wanted Children,* August, 1995

—. "NIH Violence Research Initiatives: Is Past Prologue?' Panel on NIH Research on Anti-social, Aggressive and Violence-related Behaviors and Their Consequences, September, 1993.

—. "Somatosensory Affectional Deprivation (SAD) Theory of drug and alcohol use," in *Theories On Drug Abuse: Selected Contemporary Perspectives*, 1980. Dan J. Lettieri, Mollie Sayers and Helen Wallenstien Pearson, Eds.) NIDA Research Monograph, March 30, 1980. National Institute on Drug Abuse, Department of Health and Human Services. Rockville, MD).

—. "The Origins of Human Love and Violence." *Pre-and Peri-natal Psychology Journal,* 10 (3), Spring, 1996.

—. Time Life Documentary "Rock a Bye Baby," premiered at the 1970 White House Conference On Children.

Putnam, Robert. "Bowling Alone: America's Declining Social Capital." *Current,* June, 1995.

Remland, Martin S., et. al. "Interpersonal Distance, Body Orientation, and Touch: Effects of Culture, Gender and Age. *The Journal of Social Psychology,* 1995.

Rynearson, Robert R. "Touching People." *Journal of Clinical Psychiatry,* Vol. 39, no. 6, June, 1978.

Schwade, Steve, and Linda Rao. "Hospitals with the Human Touch." *Prevention Magazine,* Vol. 46, Issue 12, December, 1994.

Seligman, Katherine. "Sex Laws For Schools." *San Francisco Examiner,* May 30, 1993

Simon, Sidney B. "Please Touch: How to Combat Skin Hunger in Our Schools." *Scholastic Teacher Magazine, Junior/Senior High Teachers Edition,* October, 1974.

Snyder, Trish. "Touch." *Chateliane,* July, 1996.

Solter, Aletha. "Television and Children." *Mothering,* Fall, 1986.

Staff. "Mothering Interviews Joseph Chilton Pearce." *Mothering,* Spring, 1985.

Stetzel, John. "Teaching Good Touch." *Joperd,* August, 1994.

Thayer, Stephen. "Encounters." *Psychology Today,* March, 1988.

Tawagoto. "Black and White," Winter, 1988, Vol.1, No.2.

Wellness Associates Journal. Issue 36, Spring, 1996.

Unsigned article. "Friendly Touch," *Human Behavior Magazine,* January, 1972.

INDEX

COPYRIGHT PERMISSIONS

Every effort has been made to trace copyright holders of the material in this book. The editor apologizes if any work has been used without permission and would be glad to be told of anyone who has not been consulted.

The lines from "since feeling is first", copyright 1926, 1954, © 1991 by the Trustees for the E.E. Cummings Trust. Copyright © 1985 by George James Firmage, from SELECTED POEMS by E.E. Cummings, Introduction & Commentary Richard S. Kennedy. Reprinted by permission of Liveright Publishing Corporation.

The lines from "somewhere I have never travelled, gladly beyond", copyright 1931, © 1959, 1991 by the Trustees for the E.E. Cummings Trust. Copyright © 1979 by George J. Firmage, from COMPLETE POEMS: 1904-1962 by E.E. Cummings, Edited by George J. Firmage. Reprinted by permission of Liveright Publishing Corporation.

Selections from: *Birth Without Violence* by Frédérick Leboyer. Reprinted with permission of the publisher, Alfred A. Knopf Inc.

Excerpts from *LOVING HANDS: The Traditional Art of Baby Massage* by Frédérick Leboyer. Copyright © 1976 by Frédérick Leboyer. ISBN 1-55704-314-0; $15.95. Reprinted by permission of Newmarket Press, 18 East 48th Street, New York, NY 10017.

ADDITIONAL HEALTH TITLES FROM HOHM PRESS

THE ALCHEMY OF LOVE AND SEX
by Lee Lozowick
Foreword by Georg Feuerstein, Ph.D., author of *Sacred Sexuality*

Discover 70 "secrets" about love, sex and relationships. Lozowick recognizes th immense conflict and confusion surrounding love and sex, and tantric spiritu practice. Preaching neither asceticism nor hedonism, he presents a middle path– one grounded in the appreciation of simple human relatedness. Topics include: what men want from women in sex, and what women want from men • th development of a passionate love affair with life • how to balance the essenti masculine and essential feminine • the dangers and possibilities of sexual Tantra the reality of a genuine, sacred marriage. . .and much more. The author is an America "Crazy Wisdom teacher" in the tradition of those whose enigmatic life and madca teaching styles have affronted the polite society of their day. Lozowick is the authc of 14 books in English and several in French and German translations only. " . attacks Western sexuality with a vengeance." —*Library Journal.*

Paper, 312 pages, $16.95 ISBN: 0-934252-58-

• • •

THE ALCHEMY OF TRANSFORMATION
by Lee Lozowick
Foreword by: Claudio Naranjo, M.D.

"I really appreciate Lee's message. The world needs to hear his God-talk. It' insightful and healing."—(John White, author, and editor, *What is Enlightenment? Exploring the Goal of the Spiritual Path.*

A concise and straightforward overview of the principles of spiritual life as develope and taught by Lee Lozowick for the past twenty years in the West. Subjects of us to seekers and serious students of any spiritual tradition include: • From self centeredness to God-centeredness • The role of a Teacher and a practice in spiritu life • The job of the community in "self"-liberation • Longing and devotion. Le Lozowick's spiritual tradition is that of the western Baul, related in teaching an spirit to the Bauls of Bengal, India. *The Alchemy of Transformation* presents hi radical, elegant and irreverent approach to human alchemical transformation.

Paper, 192 pages, $14.95 ISBN: 0-934252-62-

TO ORDER PLEASE SEE ACCOMPANYING ORDER FORM OR CALL 1-800-381-2700 TO PLACE YOUR ORDER NOW.

ADDITIONAL HEALTH TITLES FROM HOHM PRESS

'HE ART OF TOUCH: A Massage Manual For Young People
y Chia Martin

rovides young people (ages 9 and up) with a simple, step-by-step method for
:arning massage techniques to use on themselves and others for health, pain relief
nd increased self-esteem. Encourages a young person to respect his/her own body
nd the bodies of others.

'hotographs clearly demonstrate proper hand placement and capture the mood of
entleness and playfulness which the author encourages throughout the text. Adults
'ill also enjoy reading about and practicing these techniques.

'aper, 72 pages, 92 photographs, $15.95 — ISBN: 0-934252-57-2

•••

:ONSCIOUS PARENTING
y Lee Lozowick

Any individual who cares for children needs to attend to the essential message of
his book: that the first two years are the most crucial time in a child's education
nd development, and that children learn to be healthy and "whole" by living with
ealthy, whole adults. Offers practical guidance and help for anyone who wishes to
ring greater consciousness to every aspect of childraising, including: • conception,
regnancy and birth • emotional development • language usage • role modeling:
he mother's role, the father's role • the exposure to various influences • establishing
vorkable boundaries • the choices we make on behalf of our children's education ...
nd much more.

'aper, 384 pages, $17.95 — ISBN: 0-934252-67-X

TO ORDER PLEASE SEE ACCOMPANYING ORDER FORM
OR CALL 1-800-381-2700 TO PLACE YOUR ORDER NOW.

ADDITIONAL HEALTH TITLES FROM HOHM PRESS

ENNEATYPES IN PSYCHOTHERAPY
by Claudio Naranjo, M.D.

World-renowned Gestalt therapist, educator and Enneagram pioneer Dr. Claudi
Naranjo conducted the First International Symposium on the Personality Enneagram
in Pueblo Acantilado, Spain, in December 1993. This book derives from thi
conference and reflects the direct experience and lively testimony of notab
representatives of a variety of therapeutic disciplines including: psychoanalysi
Gestalt, Transactional Analysis, bodywork, and others. Each writer describes ho
the Enneagram holds invaluable keys to understanding personality and its speci
relevance to those whose task is helping others.

Paper, 160 pages, $14.95, ISBN: 0-934252-47-5

• • •

EVERYWOMAN'S BOOK OF COMMON WISDOM
by Erica Jen, Lalitha Thomas and Regina Sara Ryan

"Forget about being self-conscious. The truth is, nobody really cares. *They* are to
busy being self-conscious." So advises this postcard-sized book of bright an
profound sayings for women (and their curious male friends and spouses). Thes
sometimes provocative, often inspiring "secrets" for success in business, sex, an
family life are ideal reminders—great for posting on a refrigerator door, or sendin
in a letter to a friend. With over 100 years of combined wisdom, poet Erica Jen o
San Francisco, and Arizona authors Regina Sara Ryan (*The Wellness Workboo*
Ten Speed Press) and Lalitha Thomas (*10 Essential Herbs,* Hohm Press) offer the
unique and sometimes quirky views of life in this collection of short aphorism
about how to be happy, how to keep your man, and how to live sanely in a worl
gone mad.

Paper, 134 pages, $6.95, ISBN: 0-934252-52-1

TO ORDER PLEASE SEE ACCOMPANYING ORDER FORM
OR CALL 1-800-381-2700 TO PLACE YOUR ORDER NOW.

ADDITIONAL HEALTH TITLES FROM HOHM PRESS

'HE JUMP INTO LIFE: Moving Beyond Fear
y Arnaud Desjardins
'oreword by Richard Moss, M.D.

Say Yes to life," the author continually invites in this welcome guidebook to the spiritual ath. For anyone who has ever felt oppressed by the life-negative seriousness of religion, his book is a timely antidote. In language that translates the complex to the obvious,)esjardins applies his simple teaching of happiness and gratitude to a broad range of veighty topics, including sexuality and intimate relationships, structuring an "inner ife," the relief of suffering, and overcoming fear.

'aper, 216 pages, $12.95 — ISBN: 0-934252-42-4

•••

'OWARD THE FULLNESS OF LIFE: The Fullness of Love
y Arnaud Desjardins

Renowned French spiritual teacher, Arnaud Desjardins, offers elegant and wise counsel, arguing that a successful love relationship requires the heart of a child oined with the maturity of an adult. This book points the way to that blessed union. Topics include: happiness, marriage, absolute love and male and female energy.

'aper, 182 pages, $12.95 — ISBN: 0-934252-55-6

•••

TO ORDER PLEASE SEE ACCOMPANYING ORDER FORM OR CALL 1-800-381-2700 TO PLACE YOUR ORDER NOW.

ADDITIONAL HEALTH TITLES FROM HOHM PRES

TRANSFORMATION THROUGH INSIGHT:
Enneatypes in Clinical Practice
by Claudio Naranjo, M.D.
Foreword by Will Schutz, Ph.D.

The Enneagram, an ancient system of understanding human nature, divides huma personalities into 9 basic types (*ennea* means nine). Whether one is a newcomer t the field of Enneagram studies, or an experienced therapist using this material wi clients, Dr. Naranjo's latest book will provide a wealth of invaluable data about th Enneatypes presented in a unique format which turns a scholarly text into fascinating page-gripper. Each of the nine Enneagram types is illustrated by passage from famous pieces of literature, case studies by famous therapists, and a therapeuti dialogue between Dr. Naranjo and one of his own clients who demonstrates th type being considered. Claudio Naranjo is a world-renowned authority on th Enneagram.

Paper, 544 pages, $24.95 ISBN: 0-934252-73-

•••

WHEN HOLIDAYS ARE HELL...!
A Guide to Surviving Family Gatherings
by Mariana Caplan, M.A.

A upbeat guide for meeting the expectations and challenges of family gathering From unresolved conflicts to reconciiling differences in lifestyle, Caplan offe keys to unlocking dozens of potentially unpleasant situations.

"This book will help readers maximize the pleasures and minimize the hassle family holiday celebrations in the nerve-wracking nineties. A good job, and qui timely."—Chuck Langham, Executive Director, S.C.R.O.O.G.E. International (T Society to Curtail Ridiculous, Outrageous and Ostentatious Gift Exchanges)

Paper, 144 pages, $7.95 ISBN: 0-934252-77-

TO ORDER PLEASE SEE ACCOMPANYING ORDER FOR
OR CALL 1-800-381-2700 TO PLACE YOUR ORDER NOW.

ADDITIONAL HEALTH TITLES FROM HOHM PRESS

WHEN SONS AND DAUGHTERS CHOOSE ALTERNATIVE LIFESTYLES
by Mariana Caplan, M.A.

A guidebook for families in building workable relationships based on trust and mutual respect, despite the fears and concerns brought on by differences in lifestyle. Practical advice on what to do when sons and daughters (brothers, sisters, grandchildren...) join communes, go to gurus, follow rock bands around the country, marry outside their race or within their own gender, or embrace a religious belief that is alien to yours.

"Recommended for all public libraries."—*Library Journal.*

"Entering an arena too often marked by bitter and wounding conflict between worried parents and their adult children who are living in non-traditional communities or relationships, Mariana Caplan has produced a wise and thoughtful guide to possible reconciliation and healing...An excellent book."
—Alan F. Leveton, M.D.; Association of Family Therapists, past president

Paper, 264 pages, $14.95 ISBN: 0-934252-69-6

•••

YOUR BODY CAN TALK: How to Use Simple Muscle Testing to Learn What Your Body Knows and Needs
by Susan L. Levy, D.C. and Carol Lehr, M.A.

The World's Most Advanced Diagnostic Health Tool is at your fingertips... Your Own Body can "talk" to you, telling you what it knows and needs for health and well-being. A simple method of **energetic muscle testing** can help you to decode symptoms and become sensitive to early warnings of body dysfunction...on a daily basis—long before life-threatening illness can develop. **This book will teach you how to use energetic Muscle Testing to:** • Discover your food sensitivities • Determine effects of electromagnetic pollution in your home or workplace • Evaluate the strength of your heart, your kidneys, your liver...• Test your immune system functioning • Choose which treatment methods will best handle your condition. Special Chapters for Women cover issues of PMS and Menopause. Special Chapter for Men deals with stress and heart disease, impotence and prostate problems.

Paper, 390 pages, $19.95 ISBN: 0-934252-68-8

TO ORDER PLEASE SEE ACCOMPANYING ORDER FORM OR CALL 1-800-381-2700 TO PLACE YOUR ORDER NOW.

RETAIL ORDER FORM FOR HOHM PRESS HEALTH BOOKS

Name ______________________________ Phone (   ) ____________

Street Address or P.O. Box ______________________________________

City ____________________________ State ______ Zip Code __________

	QTY	TITLE	ITEM PRICE	TOTAL PRICE	
1		THE ALCHEMY OF LOVE AND SEX	$16.95		
2		THE ALCHEMY OF TRANSFORMATION	$14.95		
3		THE ART OF TOUCH	$15.95		
4		CONSCIOUS PARENTING	$17.95		
5		ENNEATYPES IN PSYCHOTHERAPY	$14.95		
6		EVERY WOMAN'S BOOK OF ...	$6.95		
7		THE JUMP INTO LIFE	$12.95		
8		TOWARD THE FULLNESS OF LIFE	$12.95		
9		TRANSFORMATION THROUGH INSIGHT	$24.95		
10		UNTOUCHED	$19.95		
11		WHEN HOLIDAYS ARE HELL...!	$7.95		
12		WHEN SONS & DAUGHTERS...	$14.95		
13		YOUR BODY CAN TALK	$19.95		
			SUBTOTAL:		
			SHIPPING: (see below)		
			TOTAL:		

SURFACE SHIPPING CHARGES

1st book $4.00

Each additional item $1.00

SHIP MY ORDER

☐ Surface U.S. Mail—Priority ☐ 2nd-Day Air (Mail + $5.00)

☐ UPS (Mail + $2.00) ☐ Next-Day Air (Mail + $15.00)

METHOD OF PAYMENT:

☐ Check or M.O. Payable to Hohm Press, P.O. Box 2501, Prescott, AZ 86302

☐ Call 1-800-381-2700 to place your credit card order

☐ Or call 1-520-717-1779 to fax your credit card order

☐ Information for Visa/MasterCard order only:

Card # ________ – ________ – ________ – ________ Expiration Date ______ ______

ORDER NOW!

Call 1-800-381-2700 or fax your order to 1-520-717-1779.

(Remember to include your credit card information.)